LITT

SPACE

THE LITTLE BOOK OF SPACE

Norman Ferguson has asserted his right to be identified as the author of this work in accordance with sections 77 and 78 of the Copyright, Designs and Patents Act 1988.

An Hachette UK Company
www.hachette.co.uk

Summersdale Publishers Ltd
Part of Octopus Publishing Group Limited
Carmelite House
50 Victoria Embankment
LONDON
EC4Y 0DZ
UK

www.summersdale.com

Printed and bound in Poland

ISBN: 978-1-78685-805-4

The
LITTLE BOOK OF
SPACE

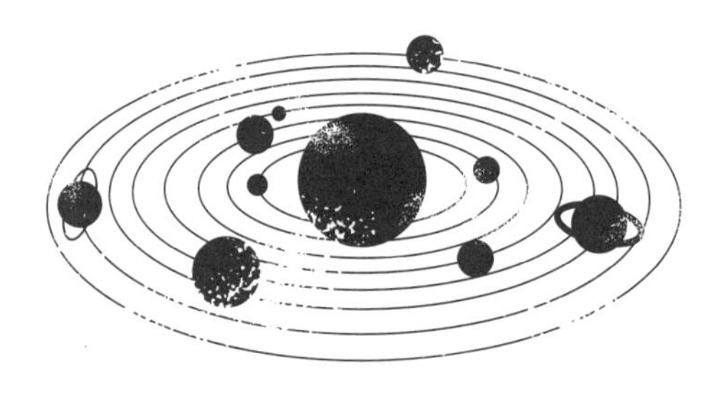

Norman Ferguson

Contents

Space is big. Really big.
You just won't believe
how vastly hugely
mind-bogglingly big it is.

DOUGLAS ADAMS,
THE HITCHHIKER'S GUIDE TO THE GALAXY

INTRODUCTION

We are all space travellers. We might not think it, but even when going to the shops, playing football or just sitting watching TV, we're racing through space. The Earth moves around the Sun and rotates on its axis extraordinarily fast. We don't notice these speeds due to gravity holding us down and everyone and everything around us moving at the same rate.

Although we are moving quickly, it still takes 365 days to go once around the Sun. Earth and the other planets form part of the Solar System, named after our own star: Sol, the Sun. As the Sun is so massive, its gravity keeps all the planets and asteroids in orbit around itself, even though the furthest objects are billions of kilometres out.

While it might seem big – and it is pretty big – the Solar System is tiny compared to the Milky Way, the collection of billions of other stars known as our galaxy, where it is located. And the Milky Way is small compared to the rest of the universe, which contains billions of galaxies. In these galaxies are many stars. It is estimated there are 70 billion trillion stars in the universe. That's a seven with 22 zeros after it, or 70,000,000,000,000,000,000,000. Words can't convey just how big it is.

Because space is so large, there is much to find out. With voyages planned to take more humans to the Moon and eventually to our nearest planetary neighbour Mars in the near future, now is the perfect time to learn all about space.

Where Does Space Begin?

There is no strict definition, but some people take space as beginning beyond the Kármán line, at 100 km (62 miles) above the Earth, while NASA and others use a lower altitude of 80 km (50 miles).

DISTANCE TO THE PLANETS

As we know, the Solar System is big. To get an idea of just how big, we can go on a journey at the speed of light, measured at 299,792 kilometres (186,282 miles) per second.

If you were in a spaceship travelling at this speed, heading outwards from the Sun, it would take just over three minutes to reach Mercury, another three minutes to reach Venus, then two and a third minutes to reach Earth. Four minutes after departing Earth's environs, Mars would be encountered.

After the four inner planets, distances and times increase markedly. We have to travel half an hour to reach the gas giant Jupiter. To get to Saturn it's another half an hour. Eighty minutes on and we are at Uranus. We arrive at the last planet, Neptune, after another hour and a half's space travel.

To give an idea how far away the stars are, it would take us over four years to get to Proxima Centauri, our nearest star.

Chapter 1

WELCOME TO SPACE

From an early age we're introduced to space, whether it's listening to the lullaby, "Twinkle, Twinkle, Little Star", or having crescent-moon lights in our bedrooms. As we grow older, we learn in school about the Solar System. As adults we are aware of coming home under a full moon and seeing sharp shadows on the ground when the sun has long set.

But how did all this magnificence start? Where did it come from? The fascinating story of how the universe of galaxies, stars and planets was created is the result of centuries of scientific observation. The path to knowledge has not always been smooth and theories remain a source of discussion and debate. For instance, one object in our night skies, Pluto, is no longer regarded as a planet, but why was it demoted?

One thing for certain is that more powerful telescopes and more advanced robotic spacecraft will only add to our knowledge about space. We can only wonder what new discoveries will be unveiled.

THE BIG BANG AND THE BEGINNING OF SPACE AS WE KNOW IT

The Big Bang is the theory widely accepted as explaining how the universe began. About 13.8 billion years ago a hot, highly condensed mass of energy called a singularity exploded, throwing out all the material that makes up our universe today. The explosion created time and space, where the universe could then exist.

At the beginning, the matter was in the form of very small particles. As it cooled, hydrogen and helium atoms were formed. Gravity caused these to come together and eventually form stars and galaxies. More elements were formed in the stars, which then produced planets and moons and, eventually, all forms of life.

Scientists believed that the universe was static until in 1924 the American astronomer Edwin Hubble discovered it was in fact expanding. Later theories emerged stating that the universe might one day begin to contract, leading to an event known as the "Big Crunch" (and then another "Big Bang"). However, the current consensus is that the expansion is actually accelerating - due to dark energy - so this theory is now thought less likely to be correct.

TIMELINE

TIME	EVENT
13.8 billion years ago	Big Bang.
13.6 billion years ago	First stars and galaxies, including the Milky Way, are formed.
4.6 billion years ago	Solar System materializes.
4.5 billion years ago	The Moon is formed when a planet-sized body collides with the Earth.
3.5 billion years ago	Life begins on Earth.
66 million years ago	Dinosaurs become extinct.
300,000 years ago	Homo sapiens (modern humans) evolve.
5,000 years ago	Pyramids are built in Egypt.
1961 CE	First human spaceflight.
1969 CE	First human landing on the Moon.

LIFE OF A STAR

Stars are born every day somewhere in the universe. They follow two main life cycles:

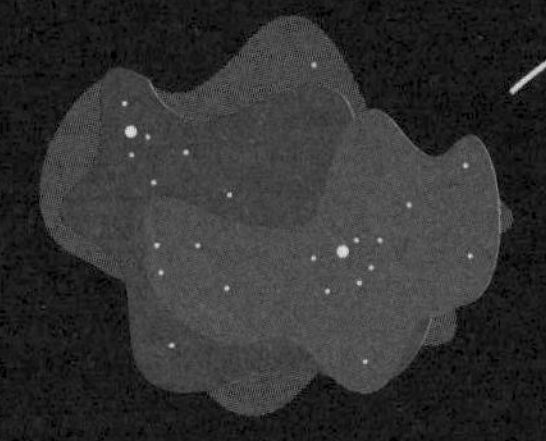

Stellar Nursery
A cloud of dust and gas contracts under gravity to form a proto-star. As it gets denser and hotter, nuclear fusion produces helium from hydrogen atoms, releasing enormous amounts of energy. The star is produced.

Main Sequence Star
Through this phase the star balances the inward pressure of gravity against the outward pressure of its nuclear fusion. Ninety per cent of stars are main sequence stars, including our Sun.

Massive Star
From the stellar nursery giant stars can form, several times bigger than our Sun.

Red Supergiant
These can form heavier elements in their core such as carbon, oxygen and iron. However, creating an iron core uses up energy. Fusion stops, and the star collapses.

Red Giant
When its hydrogen is depleted, the star's core collapses inwards. Hydrogen around the dense core is burnt and the star starts to greatly expand.

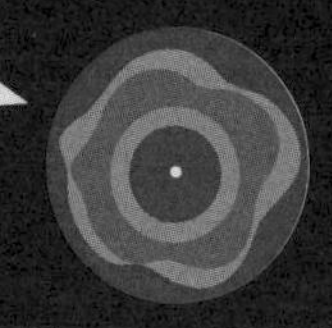

Planetary Nebula
Nearing the end of its life, the star's outer layers escape its gravity, forming a glowing cloud of gas and dust called a planetary nebula.

White Dwarf
When there is no fuel left to burn, the star's core collapses under the force of gravity to form a dense object that will cool over millions of years.

Supernova
The iron atoms being forced together produces intense heat and a shockwave; the star explodes, producing a bright supernova that can outshine its own galaxy.

Neutron Star
If the core survives it can become a neutron star, a dense object with a mass bigger than the Sun but only 20 km (12 miles) in diameter.

Black Hole
When the largest stars collapse, their cores form black holes, incredibly dense objects with a force of gravity so strong that light cannot escape.

THE MILKY WAY

The Milky Way is our galaxy, one of around 125 billion in the universe. It contains anything from one hundred to four hundred billion stars (scientists' estimates vary), our Sun being just one of them. All the stars we can see in the night sky are in the Milky Way, which until galaxies were detected in the twentieth century was thought to be the extent of the universe.

The Milky Way is vast: 100,000 light years across. Which means even if you could build a spaceship that could travel at light speed, it would take 100,000 years to make the journey.

The Milky Way is a spiral-shaped galaxy, with stars and dust and gas clouds all orbiting the centre. There are two major "arms" (Perseus and Scutum-Centaurus) and two minor arms (Norma and Sagittarius). The Sun is located in a spur called Orion, about 25,000 light years from the centre, around which it makes one complete orbit every 250 million years or so.

When you look at the stars
and the galaxy, you feel that
you are not just from any
particular piece of land but
from the Solar System.

KALPANA CHAWLA, ASTRONAUT

OUR SOLAR SYSTEM

The Solar System contains all the planets and other objects that orbit the Sun – *sol* being Latin for sun. The system consists of four inner, rocky planets (Mercury, Venus, Earth and Mars); two outer, gas giant planets (Jupiter and Saturn); two outer, ice giant planets (Uranus and Neptune); and five dwarf planets (Pluto, Ceres, Eris, Haumea and Makemake).

Between Mars and Jupiter is the Asteroid Belt, where almost two million asteroids bigger than 1 km (0.6 miles) are located. At the very far reaches of the Solar System are three huge areas: the Kuiper Belt and the Scattered Disk, consisting of millions of icy or rocky objects; and further out the Oort Cloud, containing trillions of objects, including comets. These take many years to complete their elliptical orbits, some of which are angled far above the flat plane of the Solar System. Comet C/2013 A1 Siding Spring's orbit lasts over 740,000 years, while the most well-known comet, Halley, takes a fairly rapid 76 years.

The Solar System's outer edge might be halfway to our nearest star system, Alpha Centauri.

THE SUN

The Sun is composed of gases, with three-quarters being hydrogen, the rest mostly helium, created through nuclear fusion. This fusion produces enormous amounts of energy and the Sun is therefore very, very hot. Temperatures near the surface (the photosphere) reach around 5,700°C (10,300°F), but these are dwarfed by those of the corona (the outer part of the atmosphere), which has been measured at more than 10,000,000°C (18,000,000°F). The Sun is also big: 1,400,000 km (864,000 miles) in diameter, over a hundred times that of Earth. The Sun contains 99.8 per cent of all the mass in the Solar System.

Without the Sun there would be no life on Earth, as it provides us with the heat and light essential for life to exist. This reaches us via electromagnetic radiation in the form of infrared (invisible to humans), which gives us the heat, and visible light. Also received are ultraviolet rays, which are harmful to humans; most of them are captured by the Earth's atmosphere, though the ones that do reach the surface can cause sunburn. Light from the Sun takes 8 minutes and 20 seconds to reach Earth, despite travelling at the speed of light.

MERCURY

Type: Rocky, inner
Diameter: 4,880 km (3,032 miles)
Average Distance from Sun: 58 million km (36 million miles)
Moons: 0
Name origin: Roman messenger to the gods

Mercury is the smallest planet in the Solar System, not much bigger than Earth's Moon. Despite being closest to the Sun, it is not the hottest planet; that honour belongs to Venus. During the day, temperatures on Mercury can reach 430°C (800°F) but drop to -180°C (-290°F) at night.

This small planet has an elliptical orbit, ranging 47–70 million km (29–43 million miles) from the Sun, and it's a fast one: Mercury is the speediest of planets, orbiting at over 170,000 km/h (106,000 mph). This results in an orbit of the Sun in 88 (Earth) days though its "day" – the time for it to rotate around its own axis – is 59 (Earth) days. Due to its unusual orbit and rotation, if you were on Mercury's surface the time between one sunrise and the next would be 176 days.

In appearance, Mercury is similar to the Moon, with a heavily cratered rocky surface. The biggest impact basin – a crater larger than 250 km (155 miles) across – is called Caloris, at 1,550 km (960 miles) in diameter, and was caused by the impact of an asteroid millions of years ago. Water ice has been found in deep craters on the planet's poles, probably from asteroids or comets.

Features such as craters on Mercury are named after artists, composers and writers, and include:

- Angelou (American writer Maya Angelou)
- Brontë (English writers and artists Charlotte, Emily, Anne and Branwell Brontë)
- Burns (Scottish poet Robert Burns)
- Plath (American poet Sylvia Plath)
- Tolkien (English writer J. R. R. Tolkien)
- Travers (Australian-British *Mary Poppins* writer P. L. Travers)
- Van Gogh (Dutch artist Vincent van Gogh)
- Vivaldi (Italian composer Antonio Vivaldi)

VENUS

Type: Rocky, inner
Diameter: 12,104 km (7,521 miles)
Average Distance from Sun: 108 million km (67 million miles)
Moons: 0
Name origin: Roman goddess of beauty and love

Venus is similar in size to Earth but is a much more inhospitable place. Clouds containing sulphuric acid envelop the second planet from the Sun, trapping heat and making it the hottest planet in the Solar System. Temperatures on the surface reach 471°C (880°F) – hot enough to melt lead. Coupled with this are enormous atmospheric pressures – 90 times those experienced on Earth. Venus does have an atmosphere, but it's not able to sustain human life. Carbon dioxide makes up the vast majority (96.5 per cent), with nitrogen and other trace gases making up the rest.

Measurements of the Venusian atmosphere have come from robotic spacecraft, some of which were put out of action by this hostile environment. The first landing was in 1970, when the Soviet Union's *Venera 7* transmitted information for less than an hour after

free-falling to the surface when its parachute failed. Probes have also been able to send back photographs of the rocky surface, previously hidden from view by clouds; the Soviet Union's *Venera 9* was the first to do so. Parts of this landscape are under constant change due to volcanic activity caused by the tens of thousands of volcanoes on Venus. At 8 km (5 miles), Maat Mons is the highest of these, and it also stretches 200 km (124 miles) across the planet's surface. The highest mountain is Maxwell Montes which, at 11,000 m (36,000 ft), is over 2,000 m higher than Earth's Mount Everest.

Venus rotates about its own axis in a clockwise direction, which is different to all the other planets apart from Uranus. Its rotation is also the slowest of all the planets, at 243 Earth days. As it takes 225 days to orbit the Sun, this means Venusian days are longer than its years.

EARTH

Type: Rocky, inner
Diameter: 12,742 km (7,917 miles)
Average Distance from Sun: 150 million km (93 million miles)
Moons: 1
Name origin: Germanic word for "ground"

Earth is the Solar System's only known inhabited planet and the only one to have surface water, which covers 71 per cent of its area. The deepest part of its oceans is in the Pacific's Mariana Trench at 11,034 m (36,201 ft), and its highest point is the summit of Mount Everest at 8,850 m (29,035 ft). Its land is a mixture of snow-covered poles and mountains, deserts, forests, tropical jungles and arable land with a more temperate climate.

The Earth is tilted on its axis at an angle of 23.4 degrees. This results in some areas, near the poles, having periods of no sunshine, or no night-time, depending on the season. Earth's atmosphere is mainly nitrogen (78 per cent) and oxygen (21 per cent). Its temperature provides the right conditions for many forms of carbon-based life, but the warming of the

atmosphere presents challenges to human existence. There are almost eight billion humans on Earth, sharing the planet with millions of other species, with many still to be discovered. Previously, hundreds of millions of years ago, Earth was home to dinosaurs, but these are thought to have been wiped out by an asteroid. Fears remain of a similar mass extinction event, for both humans and other species.

There is no evidence of Earth having been visited by extra-terrestrial visitors, but it has sent out spacecraft to other worlds, and several projects are seeking to determine if Earth contains the only intelligent life forms in the galaxy. In February 1990, the *Voyager 1* spacecraft was 6 billion km (3.7 billion miles) from Earth, heading out of the Solar System. It turned its camera back towards its home planet and took a photograph. This became known as the Pale Blue Dot, showing Earth at a size of one eighth of a pixel.

The Earth is a very small stage
in a vast cosmic arena.

CARL SAGAN, ASTRONOMER

MOON

Type: Rocky satellite
Diameter: 3,475 km (2,159 miles)
Average Distance from Earth: 384,400 km (238,855 miles)
Moons: 0
Name origin: Old English word for "month"

As a constant feature in our day and night skies, the Moon has attracted many myths. The exploration missions of the 1960s showed it wasn't made of green cheese, wasn't inhabited by a giant rabbit and wasn't composed of metres-deep loose dust. The Moon has been thought to influence human reproduction and moods, and the behaviour of animals. It does affect the Earth's vast bodies of water, the push and pull of its gravity producing tides.

The Moon orbits Earth every 27 days. It has no "dark side", but always presents to us the same area of its surface due to Earth's gravity slowing down its rotation. We are used to seeing the Moon move across our skies, but to someone standing on the Moon, the Earth remains in the same area of the lunar sky.

Its surface, along with hilly highland regions, has large dark-coloured areas composed of solidified lava. These are called "seas", as early astronomers thought they were covered in water. The surface is covered in craters caused by the impact of asteroids and meteorites over billions of years.

We really came to discover the Moon and yet, when we saw the Earth rise, we really discovered the Earth.

BILL ANDERS, ASTRONAUT, APOLLO 8

Tranquility Base here.
The Eagle has landed.

NEIL ARMSTRONG,
COMMANDER, APOLLO 11

Neil Armstrong (1930–2012)

Neil Armstrong became keen on aviation when taken to an air race, aged two. He gained his pilot's licence before his driving licence and built a wind tunnel to test his flying models. He flew jets in the Korean War before becoming a test pilot, flying the X-15 rocket-powered aircraft to over 63,000 m (207,000 ft) and five times the speed of sound. This height and speed he would soon surpass when he joined NASA as an astronaut. The quiet American was selected as commander of Gemini 8 and, with crewmate David Scott, was fortunate to survive when the spacecraft went dangerously out of control in Earth orbit. It was only Armstrong's cool thinking that brought the situation back under control. A reserved man who was not known for saying much, in July 1969 his words were heard around the world when announcing the first human lunar landing and then, hours later, as he stepped onto the lunar surface. Armstrong accepted his worldwide fame as the first human to walk on the Moon, but always insisted that Apollo 11's achievements were a team effort.

That's one small
step for [a] man,
one giant leap
for mankind.

NEIL ARMSTRONG,
COMMANDER, APOLLO 11

MARS

Type: Rocky, inner
Diameter: 6,791 km (4,220 miles)
Average Distance from Sun: 228 million km (142 million miles)
Moons: 2 (Phobos and Deimos)
Name origin: Roman god of war

The Red Planet is so named because of the oxidization of iron in its soil. Its dry, desert terrain is thought to have once been covered in water, and some remains both underground and in the polar regions. Mars has two ice caps at the north and south poles, both of which are composed of carbon dioxide, which freezes in winter and turns to gas in summer. There is enough ice to cover the whole planet to a depth of 35 m (115 ft). The atmosphere on Mars is thin, consisting mostly of carbon dioxide. Dust storms are regular events, lasting for several weeks, with speeds of up to 96 km/h (60 mph). Every five years or so, storms large enough to be seen from Earth envelop the planet. One such storm is thought to have halted operations of the robotic lander *Opportunity* in 2018.

Mars has several extinct volcanoes: Olympus Mons is the biggest in the Solar System, at around 21,900 m (72,000 ft) in height. Another large feature is Valles Marineris, a series of canyons ten times the size of Earth's Grand Canyon, running 4,000 km (2,500 miles) across the equator.

Over 50 robotic spacecraft have been sent to explore Mars, with several making successful landings and surface explorations. NASA's *Perseverance* touched down in 2021, carrying the first aircraft to operate from another planet, the small helicopter *Ingenuity*. *Perseverance* carried out the first successful experiment on another planet to convert carbon dioxide into oxygen. MOXIE (Mars Oxygen In-Situ Resource Utilization Experiment) demonstrated the technology that might be needed by any humans who come to inhabit Mars on a future mission.

JUPITER

Type: Gas giant, outer
Diameter: 139,822 km (86,881 miles)
Average Distance from Sun: 778 million km (483 million miles)
Moons: 79 (including the four Galilean moons: Ganymede, Calisto, Io and Europa)
Name origin: Roman king of the gods

Jupiter is the biggest of all the planets, with more than twice the mass of all the others combined. Jupiter has no solid surface, and its atmosphere is mainly hydrogen, helium and smaller amounts of methane and ammonia. Due to the high temperatures and pressures deeper inside the planet, the hydrogen is turned into a liquid, forming a large ocean, the biggest in the Solar System. Despite its size, Jupiter has the shortest day of any of the planets. It takes just 10 hours and 14 minutes for this vast world to make one complete rotation.

The gas giant is covered by bands of turbulent clouds in different shades of brown, yellow, white and orange due to their chemical composition. Its famous Great Red Spot is a perpetual storm, with speeds inside up

to 400 km/h (250 mph). It has been continuing for more than 300 years since being first observed and, although shrinking, is still larger in size than Earth.

Jupiter's four largest moons were discovered by Galileo in 1610, and this proved that objects in the Solar System did not orbit the Earth, as Christian orthodoxy stated. The biggest moon, Ganymede, is 5,270 km (3,274 miles) across and is bigger than the planet Mercury. Callisto is second-largest and, like Earth's Moon, always shows the same face to the planet it orbits; it might have an ocean deep under the surface. Io has active volcanoes while Europa is covered in water, under a crust of ice. Like Callisto, it may be an environment with the potential for hosting life. Jupiter also has several rings, but they are much fainter than those of its neighbour, Saturn, consisting of dust rather than ice.

SATURN

Type: Gas giant, outer
Diameter: 116,464 km (72,367 miles)
Average Distance from Sun: 1,426 million km (886 million miles)
Moons: 82 (including Titan, Enceladus, Phoebe)
Name origin: Roman god of agriculture, time and wealth

Often regarded as the most beautiful planet, Saturn is circled by large, highly visible rings of ice, dust and rocks – the remnants of comets and asteroids and possibly a moon torn apart by the planet's gravity. They reach out so far that, if the same rings enveloped the Earth, they would encompass the Moon. The rings are not thick, measuring an average depth of around 10 m (33 ft). A Saturn day is short, just over ten and a half hours, but its year is 29 Earth years. Saturn is almost twice as far from the Sun as its neighbour, Jupiter.

Like Jupiter, Saturn consists mainly of hydrogen and helium but has a more consistent appearance, its yellow hue the result of a hazy upper layer of ammonia crystals. It has a turbulent environment with

wind speeds reaching 1,800 km/h (1,119 mph) at the equator, four times those of the strongest hurricanes on Earth.

One feature that is common to all the giant planets is diamond rain. Methane in the atmosphere is converted into soot by powerful lightning, which then is formed into diamonds by the high pressures it encounters as it falls toward the planet.

Saturn has many moons, the biggest of which is Titan. The only satellite in the Solar System with an atmosphere, in this case predominantly made up of nitrogen, it also has surface liquids, not water but ethane and methane. Mimas has been nicknamed the "Death Star". This is due to the highly visible and very large impact crater, Herschel, which resembles a prominent feature on the planet-destroying battle station in the *Star Wars* movies, though Mimas itself wasn't photographed until the *Voyager 1* spacecraft flew past in 1980, several years after the original *Star Wars* movie had been made. Later, the *Cassini* probe discovered that another moon, Enceladus, has an underground saltwater ocean which may contain life.

URANUS

Type: Ice giant, outer
Diameter: 50,724 km (31,518 miles)
Average Distance from Sun: 2,870 million km (1,783 million miles)
Moons: 27 (including Ariel, Miranda, Titania, Oberon and Umbriel)
Name origin: Greek sky god

Discovered in 1781 by astronomer William Herschel, Uranus was at first thought to be a comet or a star. Herschel tried to name it Georgium Sidus after King George III, but this wasn't popular outside the UK and eventually Uranus was agreed on. Just as Saturn is twice as far from the Sun as Jupiter, the pattern continues with Uranus being twice as far out as Saturn. So the planet follows a long orbital path around the Sun, which takes 84 years to complete.

Uranus also has a very unusual distinction: it is the only planet on its side, being tilted at a 98-degree angle, and as such its poles spend 42 years in sunlight, then the same period in darkness. Why it is like this is not known; one theory has it that a planet-sized object smashed into it aeons ago, knocking it out of a more

conventional alignment. It also spins on its axis in a clockwise direction, sharing this feature with Venus.

The main components of the atmosphere of Uranus are hydrogen and helium, which is similar to the other giant planets. Another gas present is methane, which gives the planet its blue-green colouring, through methane absorbing the red wavelengths of the Sun's light. Despite not being the most distant from the Sun, Uranus is the coldest planet in the Solar System, with temperatures measured at −214°C (−353°F). It doesn't generate as much internal heat as the other giant planets, though the reason for this remains unknown.

Voyager 2 is the only spacecraft to have visited Uranus and only for six hours, passing by in 1986. In this time, however, it discovered new rings and moons around the planet. Uranus's rings are much darker than Saturn's, making them less visible.

Did You Know?

Uranus's moons are named after characters created by the English authors William Shakespeare and Alexander Pope.

NEPTUNE

Type: Ice giant, outer
Diameter: 49,244 km (30,599 miles)
Average Distance from Sun: 4,498 million km (2,795 million miles)
Moons: 14 (including Triton, Proteus, Nereid)
Name origin: Roman god of the sea

Neptune is so far away from Earth, it is the only planet in the Solar System that cannot be seen with the naked eye. It has twice as long a route around the Sun as its nearest neighbour, Uranus, and takes 165 years to complete an orbit, something it has only done once since being discovered in 1846, following mathematical predictions of its existence.

Being an ice giant, Neptune is mostly composed of a mix of ammonia, methane and water in a thick, liquid form, with an atmosphere of hydrogen, helium and methane, the latter of which gives it a blue colour similar to Uranus. Neptune experiences the fastest winds of all the planets, at 2,000 km/h (1,200 mph). A storm the size of Earth, dubbed the Great Dark Spot, was observed in 1989, and although it has now dissipated, others have been seen since then. Neptune

is tilted at an angle of 28 degrees, giving it seasons, albeit each one is 40 years long.

As with Uranus, Neptune has only had one visitor from Earth, *Voyager 2*, which passed by in 1989, discovering new rings and moons as it did so. It was also guided to pass by Neptune's biggest moon, Triton. First observed by William Lassell, an amateur astronomer, in 1846, soon after Neptune's discovery, Triton orbits in a different direction to the planet, the only large moon to do so. This may indicate that it was captured by Neptune's gravity and eventually – in several billion years' time – will be torn apart by this same force to become another ring.

PLUTO AND THE DWARF PLANETS

Dwarf planets are those which are generally smaller than Mercury, have enough of their own gravitational force to form into a round shape, and orbit near to other orbiting objects. There are five dwarf planets so far identified:

Pluto

In 2006 Pluto lost its status as one of the major planets, only 76 years after its discovery. Smaller than Earth's Moon at 2,302 km (1,430 miles) across, Pluto has a very elliptical orbit which means at times it is closer to the Sun than Neptune. It has five moons (Charon, Hydra, Kerberos, Nix and Styx). In 2015, *New Horizons* became the first and so far only spacecraft to visit this distant planet.

Ceres

Ceres was found between the orbits of Mars and Jupiter in the Asteroid Belt. When first discovered in 1801 it was thought to be a planet, then classified as an asteroid, but in 2006 was given the status of dwarf planet. It is only 952 km (592 miles) across.

Eris

Eris's discovery in 2003 led to a fierce debate on what constitutes a planet. This discussion led to it being named after the Greek goddess of discord and strife by the three astronomers who found it. Eris is slightly bigger than Pluto at 2,326 km (1,445 miles) across. Its orbit around the Sun is highly angled, at 47 degrees above the orbital plane of the other planets.

Haumea

Haumea is oval-shaped and spins on its axis every four hours, giving it one of the shortest days in the Solar System. It is named after the Hawaiian goddess of fertility although, when discovered in the month of December, it was initially nicknamed Santa.

Makemake

Along with Pluto, Eris and Haumea, Makemake is in the Kuiper Belt, beyond Neptune's orbit. It was discovered in 2005, and was found to be the second-brightest object in the Belt, after Pluto. It takes over three hundred years to orbit the Sun.

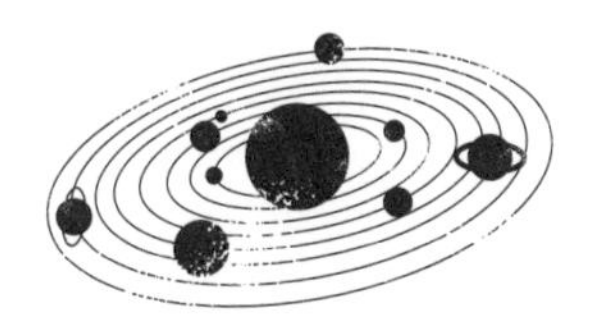

Where is
everybody?

ENRICO FERMI, PHYSICIST

LIFE IN OUR SOLAR SYSTEM – AND BEYOND

We have no way of knowing who the first human was to look at the heavens and ask the question, "Are we alone?" But the answer is still the same, that we simply don't know.

Over the centuries, writers have created alien life: H. G. Wells at the end of the nineteenth century described insect-like Selenites on the Moon and tentacled Martians invading Earth. In the 1950s Dan Dare battled the Venusian Mekon (a possible prototype for the "little green men" alien stereotype). And while the Moon, Mars and Venus might have been imagined as hosts to extra-terrestrial life, we are yet to find any confirmed evidence. But there are signs beyond our world of conditions that could support life. One promising location is Saturn's moon Enceladus, whose subsurface saltwater ocean could support life. Hydrothermal vents, similar to those found on the Earth's seafloors, could provide heat, though if life does exist there, it is likely to be microbial in form.

Beyond our Solar System, the Search for Extra-Terrestrial Intelligence (SETI) relies on the interpretation of radio signals. One project was SETI@home which used household computers to run background analysis programmes of radio signals. So far, no evidence has been found. The most intriguing signal was received in 1977 and is known as the Wow! signal, as that's what was written on a computer printout by astronomer Jerry R. Ehman when he saw unusual data coming through. But despite searching the same area of the sky on numerous occasions since then, the signal has never been seen again.

A number of theories have examined the probability of intelligent life in the cosmos. In 1961, Frank Drake's equation attempted to stimulate debate around the number of alien civilizations in the Milky Way advanced enough to send a signal detectable on Earth. The Rare Earth Hypothesis regards our planet as being the result of many fortuitous events that combined to provide complex life, which are unlikely to be replicated elsewhere. Of course, it could be that extra-terrestrial civilizations once existed elsewhere in the universe. As the introduction to *Star Wars* states, they could have existed a long time ago in a galaxy far, far away.

SETI is not science.
SETI is unquestionably a religion.

MICHAEL CRICHTON, AUTHOR

FIRST CONTACT

Television programmes and movies have shown us what an alien encounter might look like. It could be a hostile invasion, like the Cybermen in *Doctor Who*, or a peaceful visit as in *Close Encounters of the Third Kind*. There would be no way of knowing the visitors' intentions, though sci-fi writer Arthur C. Clarke thought an advanced alien civilization would have overcome their aggressive tendencies before setting off on an interstellar voyage.

Despite many claimed sightings, no conclusive evidence of any alien visit to Earth has been found, and it is likely that any first contact with extra-terrestrials will be through a radio signal. How humanity would respond would be subject to intense debate. Stephen Hawking thought it risky to contact aliens, comparing the prospect to that of Native Americans meeting Columbus.

In 2017 great interest was aroused when the first interstellar object was detected. The fast-moving, cigar-shaped 'Oumuamua was seen to accelerate away from the Sun without outgassing as a comet would. Some scientists suspected it may have been an extra-terrestrial craft.

PUTTING IT INTO PERSPECTIVE

The sheer size of space and the objects it contains can be overwhelming.

OBJECT	DIAMETER (KM)
International Space Station (ISS)	0.1
Pluto	2,302
Moon	3,475
Earth	12,742
Jupiter	139,822
Sun	1,400,000
Betelgeuse (a star)	1,063,029,600
Solar System	9,090,000,000
Tarantula Nebula	17,028,000,000,000,000
Milky Way	1,000,000,000,000,000,000
IC 1101 galaxy	4,011,040,000,000,000,000
Shapley Supercluster	1,135,200,000,000,000,000,000
Horologium-Reticulum Supercluster	5,203,000,000,000,000,000,000
Hercules-Corona Borealis Great Wall	94,600,000,000,000,000,000,000
Observable Universe	879,780,000,000,000,000,000,000

Nicolaus Copernicus (1473–1543)

Copernicus is a major figure in the history of astronomy, mainly for his assertion that the Sun was at the centre of the universe and not the Earth. This heliocentric theory had been suggested by the ancient Greek astronomer and mathematician Aristarchus of Samos some 18 centuries earlier, but his findings had not been adopted, as they went against the accepted ideas of Aristotle (fourth century BCE) and Ptolemy (second century CE), both of whom placed the Earth at the centre. This geocentric model – with the Sun, Moon, planets and stars circling the Earth – continued to dominate throughout the Middle Ages. Then in 1543 Copernicus published *On the Revolutions of the Heavenly Spheres*, which advanced the theory that all the planets, including Earth, travelled around the Sun in circular orbits. The Earth took a year to go around the Sun, said Copernicus. It also tilted as well as rotating on its own axis once each day. Of course, Copernicus's heliocentrism was so radical that decades passed before his ideas became widely accepted.

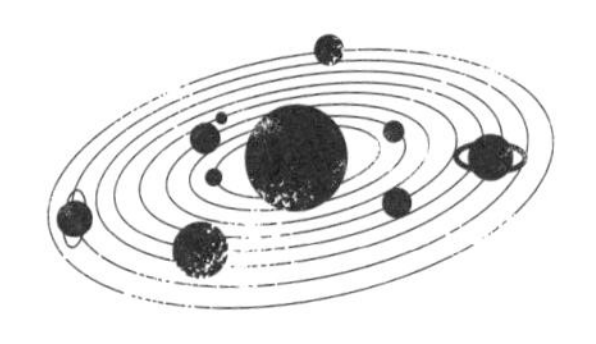

In the midst of all
dwells the Sun.

NICOLAUS COPERNICUS,
ASTRONOMER

Galileo Galilei (1564–1642)

Galileo is regarded as one of the giants of science. He advanced thinking in maths, physics and the very scientific method that is now in universal use. The Italian made some of the first astronomical observations by telescope, his many discoveries including lunar mountains and craters; Jupiter's moons; the phases of Venus; Saturn having "appendages" (rings); sunspots; and the Milky Way consisting of individual stars. Galileo's findings supported the Copernican heliocentric model. He published *Starry Messenger* in 1610 and *Dialogue on the Two Great World Systems* in 1632. His conviction that the Earth moved placed him in dispute with the rigid doctrines of the Catholic Church. Facing torture by the Inquisition, Galileo recanted his views but didn't escape being sentenced to house arrest in 1633 and his books being banned. Galileo's observations and subsequent theories led him to be called the father of modern science by Albert Einstein. In 1992, the Catholic Church finally admitted that his views on the movement of the Earth were correct.

Isaac Newton (1642–1726)

Newton was to become as influential a scientist as Galileo. The Englishman was educated at Cambridge University and later became Lucasian Professor of Mathematics there. While working at home after the plague had closed down the university, he developed many of his best ideas. One was infinitesimal calculus, but it wasn't just maths he was interested in; he carried out work in areas like theology, alchemy and physics. His research into optics demonstrated how light was composed of the "rainbow" colours of the spectrum.

In astronomy, Newton invented the first reflecting telescope and developed theories on the motions of the Moon and the planets. In 1687 he published his *Principia*, a hugely important book in the history of science. It contained his three laws of motion and his theory of gravity. The story goes that an apple falling from a tree in his garden made him wonder at the physical reasons behind its descent; "Newton's Tree" can still be seen at Woolsthorpe Manor, his home in the English county of Lincolnshire.

Chapter 2

LOOK UP: A BEGINNER'S GUIDE TO ASTRONOMY

Stargazing can be done anywhere, and you don't need any equipment to enjoy the sights of the night sky. In a city on a clear night you can easily see stars, planets and, of course, the Moon. There are many other objects we can observe from a back garden: satellites, spectacular meteor showers and the International Space Station, for example. However, to see more doesn't require much expenditure or travel. Distant nebulas or the moons around the planets of the Solar System can be viewed through binoculars or a telescope. While a visit to an observatory will provide even more sights, don't forget that we all live on our own moving observatory; as the Earth orbits the Sun we are able to see different objects throughout the year, with some constellations only visible during certain seasons.

The main thing to remember is simple: look up!

THE ORIGINS OF ASTRONOMY

Human observation of the heavens dates back millennia. This knowledge could be used for practical purposes: knowing the stars helped sailors navigate. What was in the skies also had significance for purposes we do not fully understand. Stonehenge, built 4,500 years ago, lines up with the sunrise at both midsummer and midwinter. Other sites with astronomical associations can be found all over the world. We might never know for sure what they were intended for.

Astronomy was the first science to be studied. The ancient Egyptians used stars to mark the hours of the day, and 3,000 years ago the Babylonians were making records of celestial objects, noting the movements of planets against the fixed stars. In China detailed records were kept, including the first recorded observation of Halley's Comet and that of the supernova that resulted in the Crab Nebula.

Although scientific in approach, astronomy during these early times was closer to what we would now regard as astrology, with objects and events in the skies often being used to predict events affecting humans on Earth. The Babylonians created the signs of the zodiac, dividing the sky into 12 areas.

The ancient Greeks made use of mathematics and information from the Babylonians to make great advances in astronomy. They calculated distances to the Sun and Moon, their sizes and the circumference of the Earth. They also predicted eclipses, gave names to constellations and created catalogues of the stars. Pythagoras worked out that the Earth was round after seeing its shadow on the Moon during an eclipse. In the second century CE, Ptolemy brought previous astronomical views together with his own thinking. His *Almagest* included a star catalogue with 48 constellations, and his Earth-centred view of the universe would last for almost 1,500 years.

Islamic astronomers of the period 700–1300 CE developed new ideas in maths and physics to explain the motions of heavenly objects. They gave us many of our star names, including Aldebaran, Betelgeuse and Rigel. Their insights became known in western Europe from the fifteenth century onwards. In Europe in the sixteenth and seventeenth centuries, application of the scientific method – including the advent of the telescope – provided new insights that challenged orthodox beliefs about the cosmos.

When I follow at my pleasure
the serried multitude of the stars
in their circular course, my feet
no longer touch the Earth.

PTOLEMY, ASTRONOMER

WHAT YOU'LL NEED

It is possible to see about 4,500 stars with the naked eye. Clear skies are essential, and it is best to avoid light pollution. Many more stars are visible from rural locations – sometimes so many it is difficult to make out the common constellations. Such remote locations can also provide a view of the Milky Way arching overhead. Dark sky sites offer great opportunities to see as much as possible. Stargazing on a clear night means it will probably be cold, so warm clothes, gloves and hats are advisable. Autumn, winter and spring are the best seasons, with darker and longer nights.

Binoculars or telescopes will allow you to see objects such as Saturn's rings, Jupiter's moons and distant galaxies much better; 10x50 binoculars are a good option. Telescopes vary in price but those with good light-gathering capabilities and magnifying power are preferable.

It takes about 30 minutes to acquire night vision, which allows us to see fainter objects. Take care not to ruin your night vision, especially if using star maps found on phone apps. A red torchlight can preserve night vision.

HOW TO SPOT: CONSTELLATIONS

Constellations are patterns of stars that are redolent of mythological figures, animals or objects, for example, the 13 zodiacal constellations – Taurus, Pisces, Cancer, etc. Originally named by the ancient Greeks, the constellations act as reference points in the sky. At one point there were more than 100 constellations, of which 88 are still recognized today. They include:

Ursa Major (Great Bear)

Perhaps the most recognizable. Also known as the Plough or the Big Dipper. Can be seen any time of the year in the northern part of the sky.

Ursa Minor (Little Bear)

The North pole star Polaris is the tip of the bear's tail in this smaller version of its celestial neighbour.

Cassiopeia

Shaped like a giant "W" (or "M", depending on when you see it), Cassiopeia is one of the easiest constellations to spot. It is in the northern part of the sky and so is visible all year round.

Leo (the Lion)

Best seen in spring, high in the southern sky, the zodiacal constellation Leo is easy to spot.

Taurus (the Bull)

Best seen in winter above Orion, Taurus contains a triangular-shaped centre. The bright red star Aldebaran forms the bull's right eye. Small groupings of stars like the Pleiades (Seven Sisters) are in Taurus.

Orion (the Hunter)

A landmark in the winter skies, Orion contains three stars forming Orion's Belt and the giant red star Betelgeuse.

Cygnus (the Swan)

Visible in summer, Cygnus is seen flying in front of the Milky Way.

Boötes (the Herdsman)

The Herdsman "drives" the Plough and so can be seen close by. It features the bright star Arcturus.

Auriga (the Charioteer)

Located above Taurus, this five-sided constellation featuring the bright star Capella is best seen in winter.

Pegasus

Known for its four main stars forming a square, Pegasus is a large constellation seen from autumn onwards near to Andromeda.

Gemini (the Twins)

The stars Castor and Pollux are the twins' heads and are seen right through winter, above Orion.

HOW TO SPOT: STARS

A useful one to start off with is Polaris. As the Earth rotates, the night sky revolves around this star which appears to sit above our North Pole, making it a good reference point. This also makes it a great star to aim cameras at for long-exposure star-trail photography!

Did You Know?

All stars twinkle (due to the effects of the Earth's atmosphere), but the star Algol in Perseus "blinks". It is an eclipsing binary star whose brightness reduces when the second star passes in front of it.

Top 10 Brightest Stars

The brightness of stars is measured by their level of apparent magnitude. The smaller the number the brighter the object. The table over the page shows the ten brightest that we know of.

RANKING	NAME	CONSTELLATION
1	Sirius	Canis Major
2	Arcturus	Boötes
3	Vega	Lyra
4	Capella	Auriga
5	Rigel	Orion
6	Procyon	Canis Minor
7	Betelgeuse	Orion
8	Altair	Aquila
9	Aldebaran	Taurus
10	Spica	Virgo

APPARENT MAGNITUDE	NOTES
-1.4	Also known as the Dog Star. Located close to Orion.
-0.04	Called the Bear Watcher as it is close to Ursa Major (the Great Bear).
0.03	In the movie *Contact*, extra-terrestrial signals are detected coming from Vega.
0.08	Not one star but a group of four.
0.05–0.18	Like Capella, Rigel is actually four stars.
0.34	The Winter Triangle is formed by Procyon, Sirius and Betelgeuse.
0.00–1.6	A massive red supergiant star. If at centre of Solar System would contain all the planets out to Jupiter.
0.77	With Deneb (in Cygnus) and Vega (in Lyra), Altair forms the Summer Triangle.
0.75–0.95	The brightest star in Taurus' triangle.
0.97	A binary star, forms the Spring Triangle with Arcturus and Regulus.

Jocelyn Bell Burnell (1943–)

In 1967, Jocelyn Bell Burnell was a PhD student researching quasars – very bright cores of large, distant galaxies. As part of her thesis, she helped build a radio telescope, which took the form of 120 miles of antennae in a Cambridgeshire field. Going through the results, she started to notice a repeating signal – two pulses, 1.3 seconds apart. Bell Burnell and her supervisor Tony Hewish called it "LGM-1" for "Little Green Men", as they weren't sure of its origins. Hewish thought it was radio interference from Earth, but Bell Burnell was convinced it was from space as the signals were coming from the same area of the sky. Then she found others. These newly discovered objects were neutron stars – dense, collapsed cores of supergiant stars spinning around, emitting radio waves at very regular intervals. These were named pulsars and helped to confirm Einstein's theory of general relativity and in detecting gravitational waves. When in 2018 she was awarded $3 million for her leadership in science as well as her astronomy achievements, Bell Burnell donated it to charity to fund physics students from under-represented communities.

HOW TO SPOT: PLANETS

The five planets that the naked eye can see most easily are:

Mercury - Despite its small size and being nearest to the Sun, Mercury can be spotted if conditions are clear enough. It appears after sunset low in the west or before sunrise in the east.

Venus - Venus is easily spotted in the evening or pre-dawn skies, being the second-brightest object after the Moon. It has often been mistaken for a UFO!

Mars - Mars's red colour makes it a distinctive object. It is brightest when in opposition - that is, when on the same side of the Sun as Earth - and at the closest it comes to Earth, 54.6 million km (33.9 million miles).

Jupiter - The giant planet can be brighter than any of the stars, if seen in opposition with an apparent magnitude of −2.9. Binoculars or a telescope will allow the four biggest moons and its Great Red Spot to be viewed.

Saturn - The ringed planet is easy to spot, but binoculars or a telescope will bring the rings into clear view.

HOW TO SPOT: THE MOON

With the Moon seen so regularly in both our day and night skies, we can sometimes take it for granted, but our nearest celestial neighbour deserves closer study.

It orbits the Earth once a month, visibly changing shape as it does, its phases going from a new moon to the slender crescent seen after sunset, to gibbous as it fills, to full moon, then back to a crescent (albeit with its "horns" pointing the other way). It then disappears from view before beginning the cycle again. The Moon appears to be larger when near the horizon than when higher in the sky. Despite investigation, no explanation has been given for this illusion.

The naked eye can see several lunar features such as the darker mare (the seas), the lighter highland regions and large craters such as Tycho and Copernicus. During the waxing and waning phases, as the low sun casts shadows across the lunar surface, binoculars or a telescope can give good views of the rough terrain such as the Bay of Rainbows and the Apennine Mountains.

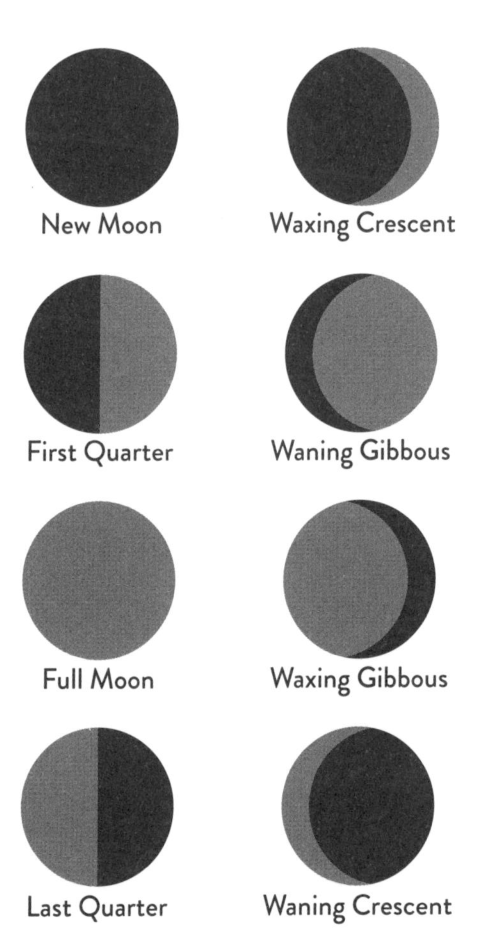
New Moon
Waxing Crescent
First Quarter
Waning Gibbous
Full Moon
Waxing Gibbous
Last Quarter
Waning Crescent

Caroline Herschel (1750–1848)

Originally from Germany, Caroline and her brother William moved to Britain to pursue careers in music, but this plan was abandoned when William followed his interest in astronomy. He gained renown for his discovery of Uranus in 1781 and was given a pension by King George III. In 1787 Caroline was similarly honoured, becoming the recipient of £50 a year – the first woman to be paid for working in the field of astronomy. She worked as her brother's assistant, writing down his observations, making calculations and preparing the telescopes for the evening sessions. When she was able to work on her own, she made several discoveries, including a series of comets. Caroline also worked on compiling several star catalogues. Her achievements were rare at a time when women's role in society was very limited. When her brother died, she moved back to Germany where, in her eighties, she worked on a catalogue of nebulae for her nephew.

HOW TO SPOT: ECLIPSES

There are two forms of eclipse, lunar and solar. A lunar eclipse occurs when the Earth passes in front of the Sun and its shadow is cast onto a full moon. The Moon can become red- or orange-tinted. A solar eclipse is much rarer and takes place when a new moon passes in front of the Sun. Quite by chance, the Moon is 400 times smaller than the Sun, which is 400 times further away, and so the Moon's disc fits perfectly over the Sun's. This spectacular cosmic event, known as totality, turns day into night for a few minutes, allowing stars and the Sun's corona to be seen. If the Moon's orbit takes it further away from the Earth, it has a smaller relative size and does not completely cover the Sun's disc, leaving a visible "ring of fire". This is known as an annular eclipse. More common is a partial eclipse, when only a portion of the Sun is obscured. Any viewing of eclipses requires special glasses or filters – you should never look directly at the Sun with the naked eye.

HOW TO SPOT: GALAXIES AND NEBULA

All the individual stars we can see are in our galaxy, the Milky Way. With dark enough conditions, the Milky Way itself is visible, its hazy, white appearance giving rise to the name.

Andromeda Galaxy

Objects beyond our galaxy can also be seen. Andromeda is our nearest galaxy, though at a distance of 2.5 million light years, we are seeing this spiral galaxy as it was 2.5 million years ago. The most distant object able to be spotted with the naked eye, it can be found close to the square of Pegasus and is visible all year round. Other galaxies that require binoculars or a telescope include the Whirlpool Galaxy, in Canes Venatici near to Boötes, and the Triangulum Galaxy in Triangulum, not far from Andromeda.

Orion Nebula

Also easy to spot are nebulae, clouds of dust and gas that can form colourful and beautiful formations. In Orion, "hanging down" from the three-star belt is Orion's sword, also comprised of three stars. The second "star" down is actually a nebula and is visible with the naked eye. It is the brightest of all the nebulae we can see.

Did You Know?

In 1504, explorer Christopher Columbus used the prediction of a lunar eclipse as a sign from God to persuade the indigenous people of Jamaica to provide his crew with food.

HOW TO SPOT: METEOR SHOWERS

As the Earth orbits the Sun, it encounters asteroids and comet trails. These burn up in the atmosphere, producing trails of light, and appear to us on the ground as shooting stars or meteors. Anything that makes it to Earth's surface is called a meteorite, around 20 of which land every day. Altogether, more than 65,000 meteorites have been found around the world. In February 2021, a fragment of a spectacular fireball was recovered from their driveway by a British family living in the Cotswolds.

Earth's encounters with comet trails are regular and expected. For example, the autumnal Orionids meteor shower is comprised of debris left by Halley's Comet. The best showers produce several meteors a minute. They can be found in the constellations from which they originate. They include:

METEOR SHOWER NAME	PERIOD VISIBLE	ORIGINATING CONSTELLATION
Quadrantids	Late December to early January	Boötes
Lyrids	Mid to late April	Lyra
Perseids	Mid-July to mid-August	Perseus
Orionids	October to early November	Orion
Leonids	Mid-November	Leo
Geminids	Early to mid-December	Gemini
Ursids	Mid-December	Ursa Minor

HOW TO SPOT: SATELLITES AND THE ISS

There around 6,000 satellites in orbit around the Earth. Some in lower orbits are easily spotted as points of light moving quickly against the stars. The best time for viewing them is shortly after sunset or before sunrise as they do not emit light but are lit by the Sun. The satellites now being produced are much smaller in size than earlier ones, meaning many can be launched into orbit in the same payload. SpaceX's Starlink project aims to have over 40,000 small satellites in operation, providing broadband coverage to those on Earth. Some have expressed reservations about the multitude of objects affecting astronomical observations.

The brightest is the International Space Station (ISS), which passes from west to east and, depending on various factors, can be visible for several minutes at a time. Its orbital path takes it over the south-west of the UK, where it can be seen directly overhead. From more northern locations it is still a prominent sight with up to five passes possible in a single night. Supply spacecraft can also be seen as they approach it. There are websites and apps which can accurately predict when it is due to pass over your location.

OTHER THINGS TO LOOK OUT FOR

Comets – Comets are rare visitors to our skies, but as they near the Sun, they can make for a spectacular sight as tails of gas and dust stream out in their wake.

Planetary Conjunctions – Conjunctions are when two or more planets are seen close together. Observing them over several nights, you will see their relative positions change.

Supermoons – A supermoon is the popular term for a full moon that occurs at the closest point in its orbit of Earth. It can be 16 per cent brighter than an average full moon.

Transits – When an object or planet crosses the face of another, larger object, it is called a transit. With proper precautions taken, Venus and Mercury can be observed as they transit the Sun.

Northern Lights – When a burst of charged particles from the Sun hits the Earth's magnetic field, it can result in a beautiful phenomenon called the Aurora Borealis. This is spotted mostly from northern Scotland upwards but occasionally can be seen at lower latitudes.

OBSERVATORIES

While older optical telescopes, housed in distinctive domed buildings, are found in our towns and cities, modern-day telescopes are generally sited away from large population centres, and at altitude, to avoid air and light pollution. Radio telescopes receive signals not as visible light but in the form of radio waves. Notable observatories include the following:

Jodrell Bank, UK

One of the first radio telescopes. Its 1957 Lovell Telescope is a Cheshire countryside landmark. It tracked early spacecraft, at one point intercepting the first images taken on the surface of the Moon, after the Soviet Union's *Luna 9* probe had landed there in January 1966.

Arecibo Observatory, Puerto Rico

This radio telescope is famous for its 305-metre-wide reflector dish built as part of the landscape in a sinkhole. It was destroyed in 2020 when a 900-tonne instrument platform fell onto the surface of the dish.

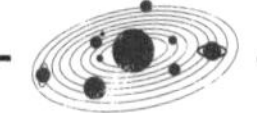

Five-hundred-metre Aperture Spherical radio Telescope (FAST), China

At 500 metres across, this is the world's largest single-dish radio telescope. Like Arecibo, it is built in a natural depression.

Mauna Kea Observatories, Hawaii

This site on a former volcano at over 4,205 m (13,796 ft) hosts 13 telescopes.

Did You Know?

The highest observatory on Earth is the University of Tokyo Atacama Observatory at 5,640 m (18,500 ft) in Chile.

TV Astronomers

In 1957 British television viewers began watching a programme on astronomy called *The Sky at Night*, presented by amateur astronomer **Patrick Moore**. He would continue presenting the show for a record 55 years. Moore combined a keen enthusiasm with an encyclopaedic knowledge of his subject. The show is now presented by **Professor Chris Lintott** and **Dr Maggie Aderin-Pocock**. Another UK scientist is **Professor Brian Cox**, a physicist who has presented programmes such as *Wonders of the Solar System* and *Wonders of the Universe*.

In America, **Carl Sagan** was an astronomer who promoted science to the public. He worked with NASA and helped provide messages to be carried on their robotic spacecraft in the 1970s. His TV series *Cosmos* is one of the most popular-ever shows about science. Sagan advocated the search for extra-terrestrial life. **Neil deGrasse Tyson** is one of the most prominent scientists in America today and director of the Hayden Planetarium in New York City. He has worked to popularize astronomy, writing numerous books and presenting radio and television shows, including a revival of Carl Sagan's *Cosmos*.

Earth is the cradle of humanity but one cannot live in the cradle forever.

KONSTANTIN TSIOLKOVSKY,
ROCKET SCIENTIST

Chapter 3

HUMANS IN SPACE

It has long been a dream of humans to journey into outer space.

Centuries before space travel became possible, writers suggested swans and large cannons as means to escape the Earth. But great scientific and engineering advances would need to be made before a human foot could step on another world.

Human spaceflight began in 1961, when astronauts and cosmonauts became real-life heroes venturing into space, strapped into colossally powerful rockets. Not all of them survived these early missions, but those that did brought back new discoveries. The reality of space travel gave those on Earth a new view of their home planet and its neighbour the Moon.

HUMAN SPACE TRAVEL TIMELINE

1957 4 October	First object in space. Soviet Union's Sputnik 1 satellite orbits the Earth.
1957 3 November	First living creature in space. A dog called Laika dies hours into the flight.
1961 12 April	First human spaceflight. Yuri Gagarin orbits the Earth for Vostok 1.
1963 16 June	First woman in space. Valentina Tereshkova completed 48 orbits for Vostok 6.
1965 18 March	First spacewalk. Cosmonaut Alexei Leonov floats for 12 minutes outside the spacecraft during the Voskhod 2 mission.
1968 24 December	First voyage to the Moon. Apollo 8 reaches lunar orbit.

1969 20 July	First human landing on the Moon. Apollo 11 astronauts take the first steps on the lunar surface.
1975 17 July	First international flight. Soviet cosmonauts and American astronauts meet in orbit as part of the Apollo–Soyuz mission.
1981 12 April	First space shuttle flight. *Columbia* reaches orbit.
1986 25 January	Shuttle disaster. *Challenger* explodes within sight of its Florida launchpad.
1991 18 May	First Briton in space. Helen Sharman travels to the Mir space station.
2020 30 May	First commercial human orbital flight. SpaceX's Crew Dragon carries two astronauts to the ISS.

Yuri Gagarin (1934–1968)

Born on a Russian farm, Gagarin became a fighter pilot before being chosen to be a cosmonaut. On the morning of 12 April 1961 at Baikonur in Kazakhstan, the 27-year-old was strapped into the tiny capsule of Vostok 1, on the top of an R-7 rocket. At 9.07 a.m. the rocket's 20 engines ignited and with a cry of "*Poyekhali*" (Let's go!) Gagarin blasted off into history. He reached a height of 327 km (203 miles) on his solo orbit of the Earth. He related what he saw below. "I see the Earth. I see the clouds. It's beautiful." Above Africa the spacecraft's retrorocket fired, slowing it for re-entry. The spacecraft went into an unexpected spin due to a still-connected cable between the re-entry module and service module but once that had burnt off, the re-entry module righted itself. Then, at a height of around 7,000 m (23,000 ft) Gagarin ejected and made the rest of his descent by parachute. He became an instant global hero, but the first human in space was denied further visits there, considered too much of an icon to be allowed back to such a high-risk environment. Tragically, he was killed in an aircraft accident in 1968.

ANIMALS IN SPACE

The first living things to be sent into space were not human astronauts but insects and animals. Animal welfare was not considered a priority and several did not survive their experience. Some of the first species in space were as follows:

Fruit flies - The first organisms in space were fruit flies launched on a V-2 rocket in 1947 from America. They reached a height of 171 km (106 miles) to test the effects of radiation. The fruit flies survived, which was a source of relief to the scientists running the operation.

Monkey - In 1949, a rhesus monkey called Albert II became the first of his kind to be launched into space, journeying aboard an American V-2. He did not survive his capsule's parachute failure on the return journey. Several subsequent Alberts also died during flights.

Mouse - In 1950 a mouse was killed following a parachute failure when returning from space in a V-2.

Dog – Laika, a Moscow stray, made the first Earth orbit by an animal in 1957, but sadly no arrangements were made for her return and she died in space.

Rabbit – Little Martha flew into space with two canine companions in 1959. All survived.

Chimpanzee – In 1961, Ham became the first chimpanzee to reach space, in a Mercury capsule being tested for human use.

Cat – Félicette, a French cat, survived her 1963 sub-orbital trip but, like several of the animals that survived their flights, was later killed for the purpose of examination.

Tortoise – In 1968, in the Soviet *Zond 5*, two tortoises circled the Moon before returning to Earth successfully seven days later.

Spider – Two spiders named Arabella and Anita spun the first webs in space in 1973 while aboard the Skylab space station.

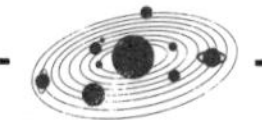

Fish – Guppy fish gave birth to live young while in orbit on board Soyuz 16 in 1974.

Other animals that have journeyed into space include jellyfish, honeybees, cockroaches, newts, frogs and a guinea pig.

THE APOLLO MISSIONS

With the Soviet Union first in putting a satellite, then a human, into space, the USA felt it had to respond. President John F. Kennedy decided a Moon landing would be the goal of a huge national effort, to be achieved by the end of the 1960s.

The project began in disaster in January 1967, when three astronauts died in a fire inside their Apollo 1 spacecraft on the launchpad. Work to ensure it couldn't happen again delayed the project but in December 1968 the giant 110-metre-high Saturn V launched three astronauts on a bold mission to the Moon. Apollo 8 orbited at Christmas, the crew taking a memorable image of the Earth as it rose above the Moon's horizon. A mission in Earth orbit tested the lunar module and then a dress rehearsal flight by Apollo 10 took astronauts down to less than 15,000 m (50,000 ft) above the lunar surface. The way was clear and in July 1969, Apollo 11 launched from Cape Kennedy in Florida in front of an estimated million people. As millions more watched and listened on television and the radio, Neil Armstrong and Buzz Aldrin descended in the *Eagle* lunar module to a safe touchdown on the surface of the Moon. Apollo 11

then returned safely to Earth. Kennedy's goal had been achieved.

Five other missions landed successfully and brought back rock samples and new discoveries about the structure of the Moon. These missions were not without incident. Apollo 12 was struck by lightning as it lifted off while Apollo 13 suffered a catastrophic explosion in one of its oxygen tanks. Rather than try and return directly, the spacecraft carried on to the Moon, using its gravity to slingshot them back home on an Earth-return trajectory. As well as the rock and soil sample collection, the astronauts also had time for comic relief. Apollo 14 commander Alan Shepard hit a golf ball with a smuggled club, and David Scott on Apollo 15 dropped a feather and a hammer at the same time to prove an experiment by Galileo that objects fall at the same rate, no matter their mass. A lunar rover vehicle was carried for the final three missions, which allowed the astronauts to travel further across the surface than before. The final mission, Apollo 17, took place in 1972, since when no human has set foot on the Moon.

Houston,
we've had a problem.

JIM LOVELL,
COMMANDER, APOLLO 13

THE FIRST HUMANS ON THE MOON

To land on the Moon presented an enormous challenge to NASA's scientists and engineers. A rocket had to be designed from scratch that was powerful enough to lift the spacecraft and its crew off the ground then send it moonwards.

NASA decided to use two spacecraft. A command module would carry the three crew members from Earth to and from the Moon, while a lunar module would take two crew members down to the lunar surface. The lunar module would then act as a launch pad for its ascent stage, which would lift the astronauts back up to rendezvous with the command module.

Four Apollo missions tried out the equipment and manoeuvres needed to navigate to the Moon before Apollo 11 put them to the ultimate test. Problems were experienced as the lunar module descended and Neil Armstrong had to fly the craft over large boulders before bringing it down safely with only seconds of fuel left.

GETTING TO AND FROM SPACE

To reach space a rocket has to attain speeds fast enough to reach Earth orbit. Those travelling to the ISS need to be going around 28,000 km/h (17,500 mph). To leave Earth's proximity and head for the Moon, Mars or anywhere else, even higher speeds are needed. Most rockets use a staging system, with each part being discarded as it uses up its fuel. Modern-day rockets produced by SpaceX are designed to return to Earth to be re-used.

It doesn't take long to reach space. In 9 minutes from blast-off a Soyuz spacecraft has reached the correct orbital speed and a height of 230 km (140 miles). From there the astronauts prepare for the final moves to rendezvous with the ISS, which orbits the Earth at a height of 400 km (250 miles).

To return to Earth, the spacecraft undocks from the ISS then fires its engine to slow it down. This causes it to start descending. As it drops towards Earth, it comes into contact with the thicker atmosphere and the outside of the spacecraft heats up to around 1,650°C (3,000°F). Inside, the astronauts are protected by a heatshield, with the cabin reaching a more pleasant 25°C (77°F). As they slow down, the

astronauts experience a pressure of up to four times the force of Earth's gravity. This can come as a shock after months of effectively weighing nothing! This re-entry part of the mission is not without risk. In 2003 the space shuttle *Columbia* broke apart when hot gases burnt into the spacecraft's wing area, which had been damaged on launch.

With the Soyuz, parachutes open at about 11,000 m (36,000 ft) to slow the spacecraft down for landing. The spacecraft, which lands on the ground, fires retrorockets just 75 cm (2ft 6 in.) above the ground to slow it down further.

SPACECRAFT

Human spaceflight requires crew-carrying vehicles and the rockets that lift them into orbit or further into space. They include:

CREWED VEHICLES

Soyuz

The Soyuz first flew in 1967 and consisted of three modules: orbital, descent and service. Cosmonaut Vladimir Komarov was killed on that test flight. Although he was alone on that trip, the Soyuz can normally carry a crew of three. The vehicle remains Russia's main spacecraft and has flown over 140 missions into space. It has continued the old Soviet tradition of landing on dry land.

Space Shuttle

This was the first reusable spacecraft, able to land on conventional runways and to carry up to eight people plus the all-important payloads, which included the Hubble Space Telescope and parts of the ISS. Between 1981 and 2011, the shuttle programme flew 135 missions, in the course of which two shuttles, *Challenger* in 1986 and *Columbia* in 2003, were lost in accidents.

SpaceX Crew Dragon

The SpaceX took its maiden flight in 2019 and was the first privately operated spacecraft to carry humans into space (up to seven at a time). As with Apollo, it lands in the sea. In May 2020, the Crew Dragon rendezvoused for the first time with the ISS, carrying two crew members, while the Cargo Dragon version continues to supply the ISS with food, equipment, clothes, etc.

ROCKETS

Saturn V

First blasting off in 1968, NASA's mighty Moon rocket consisted of three stages and had a crew of three. It weighed 2.7 million kg (6 million pounds) when fully loaded. Propelled by a mixture of liquid oxygen and kerosene, the first stage's five engines burnt fuel at a rate of 12,170 litres a second. To put its enormous power into perspective, a Saturn V produced enough energy from its first and second stages to power New York city for 1¼ hours.

Space Launch System

NASA's most powerful rocket uses two solid-rocket boosters like the space shuttle, its engines producing 4 million kg of thrust (8.8 million pounds) – 15 per cent more than Saturn V. Designed to take cargo as well as astronauts, it can carry payload of 27 metric tons (59,500 pounds) to the Moon, including the four-person Orion spacecraft. At 98 metres (322 ft) high, it is 2 metres (6 ft 6 in.) taller than Big Ben.

Eileen Collins (1956–)

Eileen Collins grew up inspired by the astronauts of the 1960s. She joined the US Air Force and went on to become a test pilot before being selected for astronaut training in 1990. Although applying to be a mission specialist, she was chosen to be a pilot. Collins flew four times into space, achieving two notable firsts.

- 1995 – First woman to pilot the space shuttle, flying *Discovery* for the first shuttle rendezvous with Russia's *Mir* space station.
- 1997 – Pilot on *Atlantis*, again to *Mir*, on a resupply mission.
- 1999 – Commander of *Columbia*; she was the first woman to lead a mission into space. Achieved successful deployment of the Chandra X-ray Observatory into orbit.
- 2005 – Commander of *Discovery* on the 'Return to Flight' first mission using the space shuttle after the *Columbia* accident. Docked with the ISS.

Collins retired from NASA in 2006.

PROBES AND LANDERS

Hundreds of probes have been launched to explore the Solar System. Among these have been some notable missions.

The Moon saw the first missions by both the USA and the Soviet Union. The Soviets' ***Luna 2*** successfully crashed into the Moon in September 1959 and a month later ***Luna 3*** photographed the far side for the first time. In January 1966, ***Luna 9*** was the first human-built object to soft-land on the Moon, with the American ***Surveyor 1*** following soon afterwards. The Soviets returned lunar soil samples on ***Luna 16*** in 1970.

Mars and Venus were explored next, with the Soviet's ***Venera 7*** landing successfully on Venus in 1970 and ***Venera 9*** sending back the first photographs of the surface in 1975. The Americans succeeded in landing two ***Viking*** landers on the Red Planet in 1976. As well as taking photographs of the surface they performed biology experiments, searching for signs of life.

A year later, ambitious explorations took place with the American ***Voyager 1*** and ***Voyager 2*** flights to both gas giants, Jupiter and Saturn. The spacecraft took advantage of a once-in-a-176-year opportunity

when the two planets aligned, using planetary gravity to redirect the spacecraft toward further targets, with ***Voyager 2*** continuing to Uranus and Neptune. ***Voyager 1*** is now the object made by humans to have gone deepest into space - over 23 billion km (14.4 billion miles) from Earth, as of October 2021.

In 1986 the European Space Agency's ***Giotto*** craft rendezvoused with Halley's Comet, getting to within 600 km (373 miles) and gaining new information on the comet's composition.

The first probe to orbit Jupiter was ***Galileo***. After being launched from the space shuttle *Atlantis* in 1989, it took six years to reach the giant planet. Having got there, it dropped a probe which transmitted data for 59 minutes before succumbing to the high Jovian atmospheric pressures. In 2003, to prevent contamination of the moon Europa, Galileo was crashed into Jupiter.

In 2001, the ***NEAR* (Near Earth Asteroid Rendezvous) *Shoemaker*** craft made the first landing on an asteroid, Eros, having orbited the 16.8-km-long object for a year before touching down.

Launched in 1997, ***Cassini***'s main objective was to study Saturn and its moons. It reached the ringed

planet through a series of flybys of Venus and Earth, gaining speed through gravity-assisted "slingshots". One of its discoveries was of saltwater ice venting into space from under the moon Enceladus's surface. *Cassini* released the ***Huygens*** probe which landed on Titan.

Another notable achievement was the return of material collected from a comet by ***Stardust*** in 2004. A sample return capsule was successfully returned to Earth two years later. Another comet encounter was enacted by ***Deep Impact***, which fired an impactor device at the comet Tempel 1 in July 2005. It impacted successfully at a relative velocity of 37,000 km/h (23,000 mph), and the debris cloud was analysed. ***Rosetta*** reached comet 67P/Churyumov-Gerasimenko in 2014 after a 10-year journey. Its *Philae* probe made the first comet landing.

The first probe to study Pluto was ***New Horizons***, launched in 2006. Despite being fast – 58,500 km/h (36,400 mph) – it still took 9.5 years to get there and, to conserve power, shut down most of its systems en route.

In recent years, a resurgence of interest in the Moon has seen probes and landers being launched by several countries, including China, Japan and India.

ROVERS AND EXPLORERS

While orbital or flyby probes provide valuable information, exploration involving robotic vehicles able to traverse the surface offers much more detailed insights on surface composition and geology. As well as scientific discoveries, they have also sent back stunning images.

The Soviet ***Lunokhod 1*** in 1970 was the first rover to explore a planetary body, travelling a distance of 10.5 km (6.5 miles) on the Moon. As the nearest rocky planet, Mars has been the focus of several missions. In 1997, appropriately on 4 July, the Americans achieved the first successful touchdown for a lander and rover (***Sojourner***) as part of the ***Mars Pathfinder*** project. For the first time airbags were used to cushion the landing. This success was followed by two **Mars Exploration** rovers: ***Spirit*** and ***Opportunity***, both in 2004. They found evidence that Mars had been covered in water. Both rovers outlasted their original scheduled service periods, with Opportunity continuing until 2019. By then it had driven 45 km (28 miles) in its 15-year-long mission. In 2012, ***Curiosity*** saw a new method of landing: being lowered to the Martian surface by a tether from a landing system that hovered using rockets. This car-sized, sophisticated rover, packed with instruments to look for signs of previous microbial life, continues its mission to the present day. More American missions have followed, including the most recent, ***Perseverance***, which landed in 2021. On board was the first aircraft to fly on another planet: the Mars helicopter ***Ingenuity***. In May 2021, China became the second country to land a rover on Mars, when ***Zhurong*** successfully touched down and began operating.

The only rovers to carry humans were the three used in the final Apollo missions. The lunar rover was able to transport two astronauts up to 7.6 km (4.7 miles) from their lunar module. The more recent increase in focus on the Moon saw the Chinese Chang'e 4 mission make the first landing on the far side in 2019, from which the ***Yutu-2*** rover was launched.

HUBBLE SPACE TELESCOPE

Telescopes located in space are above the effects of Earth's atmosphere (the reason stars appear to twinkle to us) and also the weather. There is no fog in space! There have been several space telescopes – one was used on the Moon as part of Apollo 16 – but the one launched into Earth orbit in 1990 was to have a lasting impact on our understanding of the universe.

Carried by space shuttle *Discovery*, the Hubble Space Telescope can detect infrared, visible and ultraviolet light and has made many important discoveries including evidence of dark matter, that black holes are at the centre of galaxies, and that the universe's expansion is accelerating. Among its many images are the "Pillars of Creation" of the Eagle Nebula, and 1996's Hubble Deep Field Image, which showed 1,500 galaxies in a tiny area of space equivalent in size to an 18 mm diameter coin seen from 23 m (75 ft) away. This was followed in 2004 by the Hubble Ultra-Deep Field image, containing 10,000 galaxies looking back towards the beginning of the universe.

THE JAMES WEBB SPACE TELESCOPE

The James Webb Space Telescope was launched in December 2021. It is named after the man in charge of NASA during most of the 1960s and will succeed Hubble as the main space telescope.

The telescope will be an infrared observatory, designed to look at the formation of early galaxies 13.5 billion years ago. As the universe has expanded, the light from these early galaxies is best detected via infrared, and the Webb telescope is equipped with a mirror 6.5 m (21 ft) wide for this purpose.

Hubble orbits Earth at a distance of around 547 km (340 miles) while the Webb telescope will be much further out: 1.5 million km (930,000 miles). It will be at a location called the Sun–Earth L_2 Lagrange point, where the Sun, Earth and Moon are all effectively behind it, allowing its sensors an unhindered view of the cosmos. A large sunshield is fitted to prevent the Sun's rays interfering with its sensitive equipment.

SPACE STATIONS

Space stations orbit the Earth and are used to house scientific experiments and test new technology. The first was the Soviet Union's Salyut 1 in 1971. Its first crew were unable to access the 14.6 m (48 ft)-long station, and the next three cosmonauts sent were killed when their capsule depressurized on the return journey. Three missions on NASA's Skylab (launched in 1973) overcame problems with missing solar panels and damaged heat shields to produce new solar observations and studies on humans of the effects of long-term weightlessness. Following six Salyuts, the Soviet Union began construction of the Mir in 1986. International cooperation was key to Mir's success, with a docking module being delivered by the space shuttle and American and European astronauts spending time in the station. This cooperation was continued with the ISS. China has launched its own stations: Tiangong 1 in 2011 and Tiangong 2 five years later. These have served as prototypes for a planned larger station.

Helen Sharman (1963–)

Helen Sharman was born a few weeks before the first woman flew into space in 1963. Less than three decades later, the Yorkshire woman was to follow Valentina Tereshkova into the record books.

Sharman had studied chemistry and, appropriately for someone destined to go into space, worked for a confectionery company that made a certain brand of chocolate bar named Mars. In 1989, while driving home, she heard an advert on the radio looking for British astronauts. She applied and was chosen from 13,000 applicants.

The project was a joint-effort between Britain and the Soviet Union called Project Juno, and after 18 months of training Sharman, on 18 May 1991, launched from Baikonur in Kazakhstan in a Soyuz spacecraft. Sharman spent over a week on the Mir space station where she undertook various experiments. Since her spaceflight she has spoken widely about her experiences to help inspire students in science.

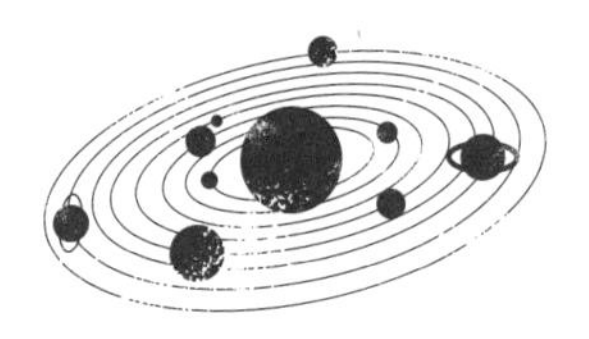

I thought only normal things could happen to me.

HELEN SHARMAN, ASTRONAUT

INTERNATIONAL SPACE STATION

The ISS is the largest human-made object in space. It is 109 m (356 ft) across – as long as a football pitch. It is of modular design with different components including crew living areas, laboratories, docking ports, airlocks and 16 large solar panels providing electrical power. One of the most popular sections is the Cupola, an observation module allowing unobstructed views of Earth. Spacewalks by astronauts helped build the station and are still performed for maintenance. Robot arms are fitted to help position new sections.

The station has been continually occupied since 2000 by over 240 people from 19 countries.

As the ISS is chiefly a research facility, it has hosted a wide range of almost 3,000 scientific experiments. Many look at the effects of long-term human exposure to weightlessness and developing technologies such as food production to aid future missions. Other areas include biotech, astronomy and Earth observation. The weightless conditions have allowed new advances to be made in vaccines, cancer treatments and metal alloys. Some experiments are mounted outside, where the vacuum of space is needed.

WHAT DO ASTRONAUTS DO?

Much of an astronaut's time is taken up with training to prepare them for future missions. It can take several years, with much to learn about the complex spacecraft and the procedures required to operate it. The ISS is made up of an American and a Russian section, so those going to live and work there have to learn to speak Russian. British astronaut Tim Peake found this one of the hardest parts of his training. After basic training, astronauts can learn how to operate the ISS's robotic arm, perform extravehicular activities (EVAs) or carry out specific experiments. Those travelling via Soyuz also have to learn how to operate the spacecraft during its launch, docking and re-entry manoeuvres.

Astronauts also have to carry out public-relations duties and, as flights approach, become more involved in media work. Another task they perform is to speak with those already up in space as the link within ground-based mission control, to pass on new information or to deal with issues as they arise.

A DAY IN THE LIFE OF AN ISS ASTRONAUT

Despite being 400 km (250 miles) above Earth and travelling at 28,800 km/h (17,900 mph), astronauts, like most of us, have routines to follow. Being weightless takes some time to get used to as there is no up or down, but new crew members soon become familiar with life in space. One effect is their feet become soft on the underside but the tops of their toes get thick and scaly as they anchor themselves to foot rails.

Typical Day

On the ISS the sun rises and sets 16 times in a 24-hour period, so to have a normal "day" crew members use Coordinated Universal Time or UTC (the same as Greenwich Mean Time). They mostly get the weekends off, although housekeeping still needs to be done.

TIME	DETAILS
6.00 a.m.	Crew members wake up. As there are no showers, they use moist towels and dry shampoo.
7.00 a.m.	Breakfast. Most meals come from vacuum packs and need warm water. Teeth cleaning is done with water and toothpaste – not rinsed out but swallowed.
7.30 a.m.	Conference with Mission Control to plan the day's work.
7.45 a.m.	Work begins. It can vary from checking experiments to taking photographs, updating social media, talking to the media, question-and-answer sessions from schools, or general maintenance. Waste is put into an uncrewed craft and then burnt up in the atmosphere.
12.00–2.00 p.m.	Lunch lasts an hour. Fresh food is much coveted whenever there's a new delivery from a cargo spacecraft. Most drinking water is recycled from sweat and from going to the toilet.
~	As prolonged periods in zero gravity affect bones, muscles and the heart, astronauts must exercise for at least two hours on a special treadmill, an exercise bike or by lifting weights.

TIME	DETAILS
7.00 p.m.	End of day conference.
8.00 p.m.	Evening meal.
~	Free time. Many choose to spend it looking out of the Cupola. They can read, send emails or speak via telephone to friends and family. Some sing songs, such as Chris Hadfield who had a hit in 2013 with David Bowie's "Space Oddity", filmed on the ISS.
10.00 p.m.	Bedtime. Astronauts have their own sleeping berths, but there are no beds. Sleeping bags prevent them from floating off as they doze. Some wear earplugs as the ISS is very noisy!

Did You Know?

As they don't get as dirty in space – and there's no washing machine – astronauts wear the same clothes, including underwear, for days at a time.

Tim Peake (1972–)

Tim Peake flew helicopters in the British Army before becoming a test pilot. He was selected as an astronaut by the European Space Agency (ESA) in 2009 and after six years of training blasted off for the ISS, the first British astronaut to do so. Principia was the name given to Peake's mission. It had a strong focus on education, and during his time in space he spoke to schoolchildren on science, engineering and on his own experiences. Peake took part in 250 experiments during his six-month stay. One involved test-driving a Mars rover located on Earth to simulate procedures on future missions. While on the ISS the former army major ran a marathon using the treadmill. He carried out an EVA, helping to replace a faulty electrical component on the solar panels and installing electrical cables. A popular and engaging communicator, Peake continues to promote STEM (Science, Technology, Engineering and Mathematics) as an ambassador.

Don't let anybody tell you you can't do anything.

TIM PEAKE, ASTRONAUT

SPACESUITS

Space is not a welcoming environment for humans. There is no oxygen to breathe, temperatures range from very low to very high, you are exposed to radiation and there is no atmospheric pressure. If you were to step outside your spacecraft, you would have seconds before your body's fluids boiled. And there is always the chance of being hit by a micrometeoroid.

To protect themselves, astronauts wear spacesuits. Backpacks carry oxygen. Internal temperatures are carefully controlled by circulating cool water. To keep fingers warm, electrical heaters are fitted in the fingertips of gloves. Gold-plated helmet visors reflect the Sun's dangerous rays. The suit must be strong enough to retain the correct atmospheric pressure but flexible enough to allow enough movement. On the first spacewalk, in 1965, Alexei Leonov's suit ballooned and he struggled to get back into his spacecraft. In 2013 Italian astronaut Luca Parmitano faced another hazard when his helmet filled with water and he had to quickly return inside.

Humans in Space Records

Longest Spacewalk – In 2001, James Voss and Susan Helms spent 8 hours and 56 minutes on a spacewalk on the ISS.

Most Spacewalks – Russian cosmonaut Anatoly Solovyev has performed 16 spacewalks.

Furthest Humans Have Been from Earth – The crew of Apollo 13 reached a distance of 400,171 km (248,655 miles) from their home planet as they swung around the Moon on their return journey.

Most Time Spent in Space – Russian cosmonaut Gennady Padalka has spent 878 days in space during five flights.

Oldest Person in Space – *Star Trek* actor William Shatner was 90 when he flew on Blue Origin's New Shepard in October 2021.

Most Isolated Person – In 1984, Bruce McCandless test-flew the untethered Manned Manoeuvring Unit a hundred metres away from the space shuttle *Challenger*.

Valentina Tereshkova (1937–)

Valentina Tereshkova worked in a textile factory in Russia before volunteering for the Soviet space programme. Although not a pilot, her passion and ability for parachuting meant she was accepted. After 18 months of training, she launched into space on Vostok 6 on 16 June 1963, becoming the first woman in space. Her mission lasted three days, during which she orbited the Earth 48 times. However, it almost ended in tragedy due to an error in the ship's navigation system. If Tereshkova had not noticed the mistake, she would have flown out into space rather than back to Earth. Once the correct data was sent up, her superiors made her swear to secrecy, and this detail was only uncovered 30 years later. After re-entry she landed on the China–Russia border, where villagers helped her out of her spacecraft and invited her for dinner. Tereshkova was an inspiration and role model to young women everywhere, but particularly to aspiring astronauts.

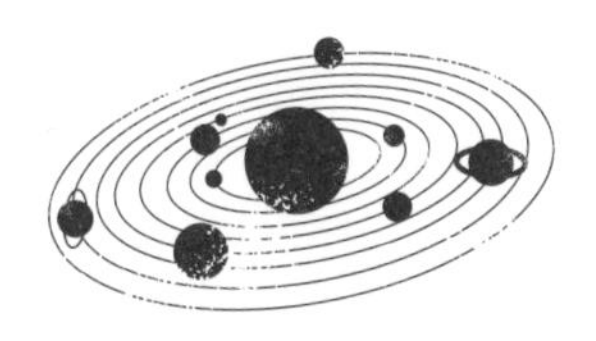

If women can be railroad
workers in Russia,
why can't they fly in space?

VALENTINA TERESHKOVA,
COSMONAUT

SPACE JUNK

One of the unfortunate results of spaceflight is the amount of debris left in Earth's orbit. Pieces of spacecraft, abandoned satellites, mislaid tools – all continue to circle the planet. While space is big, there is still a chance of impact, and with orbital speeds of around 8 km (5 miles) a second, even the smallest object would have a devastating effect.

There is a lot of it: around 3,000 defunct satellites and over a 100 million smaller fragments bigger than one millimetre in size. Thousands of these pieces originated from military missions testing anti-satellite weapons. An explosion in space throws these fragments out in a large cloud of debris.

The US Space Surveillance Network tracks over 27,000 larger objects and, to avoid collisions, the ISS can change its orbit. While the chances of collision are small, they are real. In 2009 an American satellite was destroyed by a collision with a Russian one over Siberia, and in 2021 a hole was made in the ISS's Canadarm2 robotic arm.

Dorothy Vaughan, Katherine Johnson and Mary Jackson

The 2016 Hollywood movie *Hidden Figures* tells the story of three African-American women who worked in the early days of America's space programme. **Dorothy Vaughan** (1910–2008) began working on aeronautical research data during the Second World War and was promoted to become the first African-American manager in the National Advisory Committee for Aeronautics (NACA), the predecessor of NASA. She was in charge of the West Area Computing team of human "computers", i.e. mathematicians, at Langley Research Centre.

One of these was **Katherine Johnson** (1918–2020) who worked on orbital mechanics, calculating the paths of the spacecraft. When NASA launched its first astronaut, Alan Shepard, in 1961, Johnson analysed his Mercury rocket's trajectory, and when John Glenn was about to launch on his three-orbit flight he specifically asked for Johnson to verify the computer-calculated orbit. She subsequently worked on the Apollo and space shuttle programmes.

Mary Jackson (1921–2005) also began as a computer at Langley under Vaughan but then trained as an engineer, becoming NASA's first African-American female engineer. In 2020 NASA's Washington headquarters was named in her honour.

Scientists said, "What good does it do us to go to space?" Well, what good does it do to stay home?

KATHERINE JOHNSON

Chapter 4

THE NEW SPACE RACE

After eight decades of exploration, space remains a fascinating area of research. There are no signs of interest diminishing in the future – if anything it's increasing. Private companies and nations like China and India have launched their own space rockets and are planning missions to the Moon, some of which will carry humans.

Space is no longer seen just as a place to visit. Thoughts have turned to securing a new home for humanity if Earth becomes uninhabitable. This presents enormous challenges. How can we live and work in space? What sorts of worlds will humans look for?

SPACE TOURISM

While Spain, Bali or the Outer Hebrides might be the holiday destinations of choice for some, for others there is only one place to spend their vacation: outer space.

In 2001, American businessman Dennis Tito became the first space tourist when he paid $20 million to visit the ISS for seven days, flying there in a Soyuz. He was followed by seven other individuals including the first woman space tourist, Anousheh Ansari, whose family were behind the Ansari X Prize to reward the first reusable low-cost space vehicle.

The concept of paying for a space journey goes back to the 1960s when, following its depiction in the film *2001: A Space Odyssey*, American airline Pan Am started collecting names of those wishing to travel to the Moon. Now, in the twenty-first century several private companies are offering trips to space.

Virgin Galactic

Richard Branson's company are offering sub-orbital flights in SpaceShipTwo craft. Launched from a carrier aircraft at 15,000 m (50,000 ft), it aims to carry eight passengers to heights around 100 km

(62 miles), where they can experience weightlessness for a few minutes, before descending to a landing on a runway. The cost is $450,000 a seat. The first passenger-carrying flight took place in July 2021.

Blue Origin

Owned by Amazon founder Jeff Bezos, the Blue Origin's New Shepard rocket will also take passengers on sub-orbital flights in a fully autonomous capsule (with no pilot on board). It also plans to take paying passengers to the Moon. The cost of a flight is around $200,000. It also made its first passenger-carrying flight in July 2021.

SpaceX

SpaceX was created by another billionaire, Elon Musk, with a long-term goal of colonizing Mars. SpaceX has flown astronauts to the ISS and plans to take private citizens there – for $50 million. Beyond that, the first commercial passenger flights around the Moon are planned, with Japanese billionaire Yusaku Maezawa buying all eight seats for a five-day flight aimed for 2023.

At the moment the only habitable destination is the ISS, though American hotel chain owner Robert Bigelow has also developed the concept of the space hotel, using inflatable sections. One test section was installed on the ISS in 2016, but the company ran into financial difficulties in 2020 and the project is now on hold.

THE FUTURE FOR NASA

Formed in 1958 in response to the Soviet Union's Sputnik 1 mission, NASA remains America's government-funded space agency. The space shuttle's last flight in 2011 did not mean the end of NASA astronauts flying to space, but they now had to go via Russian Soyuz spacecraft. However, with private companies now operational, launches from American soil have returned.

The agency plans to put humans back on the Moon by 2025 as part of the Artemis programme, one of its goals being to land the first woman on the lunar surface. NASA has also developed its own space vehicles. The Orion spacecraft, launched by the powerful Space Launch System rocket, would use the Gateway space dock in lunar orbit as a transfer facility for lunar landers. SpaceX have been contracted to use their Starship HLS spacecraft. Looking like a rocket from a 1950s sci-fi movie, the 50 m (164 ft)-tall ship will descend to the lunar surface then return the astronauts to Gateway. NASA plans to use its experience from Moon landings to land the first humans on Mars.

MISSION TO MARS

It took three days for astronauts to get to the Moon. At less than 400,000 km (249,000 miles) away, in space terms it's relatively close. The next planetary body to be visited is our nearest planet: Mars. The Red Planet is much further away. Even when Earth and Mars are at their closest point, Mars is still 54.6 million km (33.9 million miles) away – over 130 times the distance to the Moon.

The long distances mean any human mission will involve years in space. It will take around nine months to get there, then a stay of three months until the two planets come near to each other again, before a launch from Mars and another nine-month-long return journey to Earth. This might seem daunting, though at least there is a return journey. In 2012, the Mars One project proposed sending humans to Mars with one condition: they wouldn't be coming back; several thousand people signed up but the company went bankrupt in 2019.

There is no doubting that the long duration of a mission to our neighbouring planet presents huge issues to be overcome, but if successful, a crewed landing on Mars would mark another major landmark in human history.

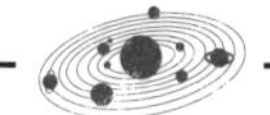

I am convinced that humans need to leave Earth and make a new home on another planet. To stay risks annihilation.

STEPHEN HAWKING, PHYSICIST

THE HUNT FOR NEW HOMES

Finding a new home is an insurance policy in case anything happens to life on Earth. Humans face mass extinction threats from themselves through war or climate change, or from space in the form of asteroid impact. But where can humankind live?

Mars

One of the reasons for journeying to Mars is to see whether it can provide a future home for humans. Those wishing to colonize the Red Planet face many physical challenges that might appear insurmountable: dust storms, radiation unfiltered by an Earth-like atmosphere, no organic soil to grow crops, no drinking water and, of course, a lack of oxygen to breathe. However, in 2021 an experiment by the *Perseverance* rover produced oxygen from carbon dioxide found in the atmosphere. This could pave the way for oxygen being produced for breathing as well as for powering returning spacecraft.

Exoplanets

If humans venture further beyond the Solar System, they will look for planets with similar characteristics to Earth. Planets that are similar to Earth are known as exoplanets, found around stars in what is called the "Goldilocks Zone". The planet's orbit mustn't be too close to the star, where it's too hot and surface water essential for life evaporates, or too far away, where it's too cold and water turns to ice. The planet would need to have the same strength of gravity as our home planet and no poisonous gases in the atmosphere, of course!

Getting There

There could be 500 million habitable planets in our galaxy, and finding one with the right conditions represent the first steps taken by humans in finding a new Earth. One candidate is Proxima b, a planet in the right area orbiting Earth's nearest star, Proxima Centauri. In space all things are relative, as this still means it is 40 trillion km (25 trillion miles) away.

In sci-fi movies, hyper drive or warp drives are used to make interstellar travel possible. Wormholes are another convenient way of quickly getting from star system A to star system B, but these ideas are still fictions, and without them it will take a long time. The fastest human-carrying spacecraft so far was the command module from Apollo 10. At its maximum speed of 40,000 km/h (25,000 mph), it would take around 115,000 years to reach Proxima b. The fastest-ever spacecraft, the Parker Solar Probe, is expected to reach a speed of 724,000 km/h (450,000 mph). It would still take over 6,000 years to reach the planet.

Even with faster spaceships, the enormous length of time means humans starting the journey would not finish it; their descendants would be the first to set foot on a new planet. Or it could be robots, perhaps built on 3D printers along the way.

There is another consequence of high-speed space travel. Einstein outlined in his theory of relativity that as you near the speed of light, time passes at a different rate than for those who remain stationary. An astronaut travelling at such high speeds for 10 years might come back to discover 100 years have passed on her home planet Earth.

OFF-WORLD RESOURCES

In the 2009 film *Moon*, Earth's satellite is mined for helium-3 – the solution to the planet's energy problems. The mineral, not found on Earth, produces radioactive-free nuclear fusion. One reason for mining the real-world Moon is to acquire water found at the south pole, which could be used for rocket fuel. Asteroids have the potential to be mined for rare elements such as platinum and gold. One asteroid, 511 Davida, is estimated to be worth $100 trillion.

There are two issues concerning off-Earth mining. One is how to do it. By necessity, mining operations would have to be carried out a long distance away from Earth in a hostile environment. A mining craft would have to land on a gravity-free asteroid in the Asteroid Belt and then return the material to Earth. Material could be worked on in space which entails the equipment and facilities needing to be launched along with the mining craft itself. The other issue is one of ethics. Some feel it isn't justifiable to carry out destructive space mining in the same fashion as happens on Earth.

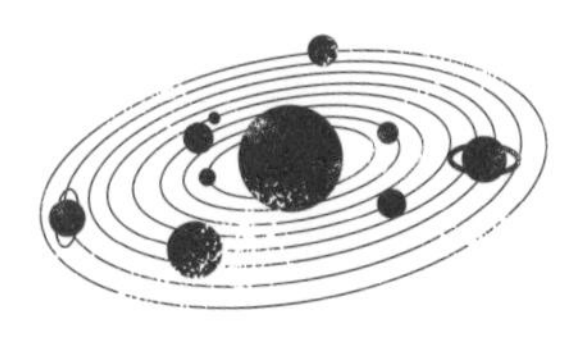

The first trillionaires will be those who mine asteroids.

NEIL DEGRASSE TYSON,
ASTROPHYSICIST

WAR IN SPACE

The Space Race between America and the Soviet Union took place during the Cold War when both superpowers competed for supremacy without any actual fighting. It began following the flight of Sputnik 1 in 1957 when the Soviets demonstrated they could launch objects that could overfly any country. Americans were concerned bombs could be launched in the same way. International treaties stated that outer space should only be used for peaceful reasons, but the military have been able to use space for various purposes.

Reconnaissance satellites use long-lens cameras to spy on opponents and the GPS (Global Positioning System) satellites – as well as guiding our car satnavs – give accurate target-location information for delivering weapons. Some space shuttle missions were flown for military purposes, with details of the payload they carried remaining secret. The Soviets used the Salyut space stations for military reconnaissance purposes. Methods for destroying satellites have been developed and in 1985 an American F-15 fighter jet fired a missile that destroyed an orbiting satellite. China has also developed anti-satellite technology, though the result of any such destructive act is yet more space debris circling the Earth.

In the 1980s US President Ronald Reagan announced the Strategic Defense Initiative – nicknamed "Star Wars" – to create a network able to shoot down any incoming ballistic nuclear missiles. However, SDI was never realized. The end of the Cold War in the early 1990s meant the US and the former Soviet Union could now co-operate more in space for peaceful ends, such as on the ISS. But new tensions in recent years between these two world powers may well thwart further collaboration.

Did You Know?

In the 1950s, the Americans and Soviets both considered detonating a nuclear weapon on the Moon as a sign of technological prowess, but instead decided to pursue its peaceful exploration.

THE END OF THE WORLD

Many have wondered how the world will end. Science fiction writers and film-makers have created their own depictions, with some more likely to occur in the real world than others.

Bypass

In the sci-fi book *The Hitchhiker's Guide to the Galaxy*, written by Douglas Adams, the Earth is destroyed by aliens building a "hyperspatial express route".

Planetary Collision

In the 2011 film *Melancholia*, a planet collides with Earth, an outcome that's not too far from plausible reality. It is thought a large planetary body colliding with the fledgling Earth created the Moon billions of years ago.

Asteroid Impact

The potential of a large asteroid or comet heading for Earth has been portrayed in movies like *Deep Impact* (comet) and *Armageddon* (asteroid). There are precedents for such an event. Scientists believe that 66 million years ago a mountain-sized meteorite crashed into Mexico's Yucatán Peninsula. The debris thrown into the atmosphere blocked out much of the sunlight, damaging plant life and compromising the food chain. An estimated 75 per cent of all animal and plant species and 100 per cent of all the non-avian dinosaurs became extinct. In 1908 over 2,000 square km (772 square miles) of Siberian forest was flattened by an exploding meteor and in 2013 a 10,000-tonne asteroid exploding above the Russian city of Chelyabinsk shattered windows, injuring over a thousand people.

Nuclear Armageddon

An exchange of nuclear weapons was a common fear in the Cold War years. It was called Mutually Assured Destruction as the resultant radioactivity would make much of the world uninhabitable. While the Cold War has ended, large stockpiles of nuclear weapons are still held by major powers.

Climate Change

While humans and other species might perish or suffer large-scale extinctions, as has happened several times in the past, the planet itself will endure.

Solar Expansion

The real end of the world will be dramatic but will take place over a long timescale. At the end of its life, when the Sun has exhausted all its fuel and become a red giant, its outer layers will expand so far out from the core that they will engulf the Earth. But this is not due to take place for another 5 billion years, so there is no need to panic just yet.

Stephen Hawking (1942–2018)

While a student at Cambridge, Hawking was diagnosed with motor neurone disease, a degenerative, muscle-wasting condition, and was given only a few short years to live. Despite this gloomy prognosis, he continued his work on understanding the nature of the universe, something he pursued throughout what turned out to be a long and very full life. Hawking spent his career at Cambridge University, where in 1979 he became Lucasian Professor of Mathematics, a post held by Isaac Newton three centuries earlier. Among his achievements was the theory of Hawking radiation, that black holes emit radiation.

In 1985, Hawking lost his voice after a tracheostomy and thereafter used a distinctive voice synthesizer to communicate. But he was still keen to make his research and knowledge accessible to the general public. In 1988, his book *A Brief History of Time: From the Big Bang to Black Holes* became a bestseller, selling over 20 million copies, although not all who bought it understood its contents fully.

Elon Musk (1971–)

Elon Musk is one of the world's richest people, having an estimated fortune of $150 billion.

He was born in South Africa and grew up in Canada before becoming a tech entrepreneur in the USA. He co-founded several companies, one of which included PayPal, making millions from their subsequent sales. In 2004, Musk joined electric car-making firm Tesla as chairman, becoming chief executive four years later.

In 2002, Musk used the wealth he had amassed by then to start SpaceX. The company's goal is to produce reusable spacecraft for the purpose of travelling to Mars. Its first rocket, Falcon 1, launched in 2008. At times, Musk has courted controversy but he remains an important figure in the new era of space travel.

Did You Know?

SpaceX launched a car into space. In 2018, on its first flight, the Falcon Heavy rocket lifted a Tesla Roadster, with a dummy astronaut called Starman in the driving seat. Driver and vehicle are now in orbit around the Sun.

SPACE MYTHS

Urban myths are something we are all used to hearing. Do alligators really live in New York sewers? Space has acquired its own set of myths that aren't actually true.

Space pens – The story goes that the Americans spent millions developing a type of pen able to be used in space while the Soviets just used pencils. In fact, the pens were developed by a private company and then sold to NASA for $2.95 each.

Moon landings – Despite the overwhelming evidence of the landings, including film footage, camera still images, personal testimony and radar tracking, some believe they never took place and were filmed on a movie set.

The Earth is flat – A surprisingly large number of people believe that the Earth is not a spherical object and also that it sits at the centre of the Solar System.

A face on Mars – In 1976 the *Viking 1* probe sent back a photo of the surface of Mars that included a feature resembling a human face, possibly a remnant of an alien civilization. Subsequent images taken by later probes showed it was just a geographical feature called a mesa.

The sight of the stars
always makes me dream.

VINCENT VAN GOGH, ARTIST

SPACE IN THE ARTS

Space has been a continuing source of inspiration for artists, composers and writers. Some notable works are as follows:

ART

Starry Night over the Rhône (1888)
and *Starry Night* (1889)
Dutch post-Impressionist artist Vincent van Gogh painted two landscapes featuring vivid representations of the night sky.

Museum of the Moon (2016)
A 7 m-wide 3D representation of the Moon created by Luke Jerram has been exhibited around the world.

The Planets (1914–16)
Gustav Holst
Holst composed a movement for each planet, expressing their astrological character, e.g. Mars is the "Bringer of War".

"Space Oddity" (1969)
David Bowie
Bowie sings about astronaut Major Tom in this single released days before Apollo 11 launched for the Moon.

"'39" (1975)
Queen
Guitarist and astronomy enthusiast Brian May sang this song about a crew returning from a voyage to find much time has passed on Earth.

Doctor Who
The Time Lord has traversed both time and space in the famous TARDIS police box since 1963.

Star Trek
There have been several spin-off shows and films in the Star Trek universe since it first ran as a TV series in 1966.

Space: 1999
The Moon is knocked out of its orbit, taking the occupants of Moonbase Alpha on an extraordinary journey into interstellar space. The series ran from 1977 to 1979.

FILMS

2001: A Space Odyssey (1968)
Monoliths are discovered on Earth and the Moon, placed there by a superior alien intelligence.

Arrival (2016)
After 12 large alien spacecraft arrive around the globe, a team attempt to communicate with the aliens.

The Martian (2015)
A stranded astronaut on Mars has to improvise in order to survive.

BOOKS

Jules Verne
From the Earth to the Moon (1865)
Humans travel to the Moon in a capsule fired from a cannon.

H. G. Wells
War of the Worlds (1897)
Martians invade Victorian England in one of the first sci-fi novels.

Tom Wolfe
The Right Stuff (1979)
Wolfe's classic book describes the tension between test pilots and Mercury astronauts in the early US space programme.

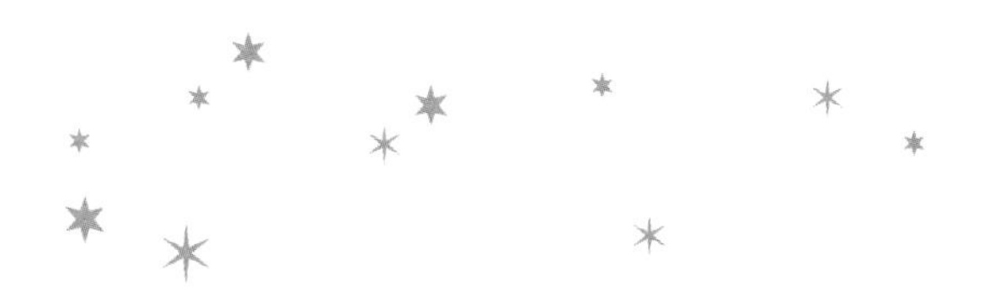

CONCLUSION

It's an exciting time to be interested in space. New discoveries are being made all the time, with human spaceflight missions from several nations planned in the years ahead. Within ten years, fresh human footprints will have been left on the Moon and new ones on Mars. Will those on the Red Planet help answer one of the biggest mysteries: is there life on Mars?

Sometimes the results of space missions are unexpected, as with the view of Earth seen for the first time by the Moon-bound Apollo astronauts, or from *Voyager 1*, showing our planet as a tiny part of the giant cosmos. We can only wonder at our own reactions on seeing images of Earth taken by the first explorers on Mars as they look up from that desert planet.

While there is much to look forward to, space and its exploration does not come without difficult questions. Many query the cost of these hugely expensive missions and that of sending humans up for a short hop into space. Would this money be better spent elsewhere? However, as was noted around the time of Apollo, not a single dollar of that mammoth project's budget was spent on the Moon. It all went to companies and employees on Earth.

There is also risk. Spaceflight does not come without danger. Several crew members have died journeying to or returning from space, and as more and more take to the stars, the risks remain. Riding into orbit in a metal tube propelled by hundreds of tons of burning liquid fuel is not near the top of the list of safest activities. Is it worth it?

Space has always provided excitement and thrills as humans reach out from their home planet. No one who was around at the time of the first Moon landing will forget the global sense of occasion and we can only anticipate what exhilarating moments await us in the years to come.

The conquest of space
is worth the risk of life.

GUS GRISSOM,
ASTRONAUT

GLOSSARY

Asteroid – A rocky minor planet.

Black hole – A dense collapsed star whose gravitational pull is so strong that even light cannot escape.

Comet – Accumulations of ice and rocks that when orbiting the Sun produce long tails through heating. There are over 3,700 identified comets.

Constellation – A grouping of stars forming the basis for a recognized shape or structure.

Dark energy – This theoretical force makes up around 68 per cent of the universe and is thought to be the reason the universe's rate of expansion is accelerating.

Dark matter – Makes up around 27 per cent of the universe. It has not been directly detected but is measured through gravitational effects.

Exoplanet – A planet found outside the Solar System.

Light year - The distance covered if travelling at the speed of light for 365 days: around 9 trillion (9,460,000,000,000) km.

Meteor - A meteoroid that glows as it encounters Earth's upper atmosphere.

Meteor shower - When Earth passes through the trail of a comet, large numbers of meteors can be seen. Well-known showers include the Lyrids, Perseids and Leonids.

Meteoroid - Rocky fragments of comets or asteroids.

Meteorite - A meteoroid that survives entry into Earth's atmosphere to fall onto the planet's surface.

Pulsar - A spinning star emitting radio waves at regular intervals.

Quasar - A bright object at the centre of a galaxy, fuelled by dust and gas acquired by massive black holes.

RESOURCES

WEBSITES

Space.com
www.space.com

NASA
www.nasa.gov

National Space Centre
www.spacecentre.co.uk

Heaven's Above
www.heavens-above.com

Astronomy Picture of the Day
https://apod.nasa.gov/apod/archivepix.html

BOOKS

Piers Bizony, ***The Search for Aliens: A Rough Guide to Life on Other Worlds*** (Rough Guide, 2012)

Andrew Chaikin, ***A Man on the Moon: The Voyages of the Apollo Astronauts*** (Penguin, 2019)

Heather Couper and Nigel Henbest, ***The Story of Astronomy: How the Universe Revealed Its Secrets*** (Cassell, 2012)

Robert Dinwiddie, ***A Little Course in Astronomy*** (Dorling Kindersley, 2013)

Tim Peake, ***Ask an Astronaut*** (Century, 2017)

Giles Sparrow, ***Spaceflight: The Complete Story from Sputnik to Shuttle and Beyond*** (Dorling Kindersley, 2007)

THE LITTLE BOOK OF MANIFESTATION

Astrid Carvel

£6.99
Paperback
978-1-80007-262-6

IT'S TIME TO START CHANGING YOUR LIFE

The universe is ready to give you what you want, if you're willing to make the right moves. Whether you dream of improving your health, career, finances or love life, the practice of manifestation may just hold the key. This beginner's guide is here to explain what manifesting is, how it works, and the simple steps you can take to get started and keep going. Bring your aspirations into being as you learn how to wield the power of positive energy.

THE LITTLE BOOK OF WORLD MYTHOLOGY

Hannah Bowstead

£6.99
Paperback
978-1-80007-176-6

STEP INTO A WORLD OF GODS, HEROES AND MONSTERS

Mythologies have been fundamental to cultures and societies throughout history and across the world. This pocket guide offers the perfect introduction to the major world mythologies, exploring their origins, foundational stories and key mythological figures. If you're looking to enrich and expand on your understanding of world history, religion and culture, then this book is an ideal starting point to fill your mind with stories of wisdom and wonder.

Have you enjoyed this book? If so, find us on Facebook at **Summersdale Publishers**, on Twitter at **@Summersdale** and on Instagram at **@summersdalebooks** and get in touch. We'd love to hear from you!

www.summersdale.com

Image credits

Cover background © Vegorus/Shutterstock.com
Cover, p.3 and throughout © Meowu/Shutterstock.com
pp.14–15 © Marusya Chaika/Shutterstock.com
pp.20, 22, 24, 27, 32, 34, 36, 38, 40 © Daryaart9/Shutterstock.com
p.69 © Ice_AisberG/Shutterstock.com
pp.62, 72–73, 89, 98, 122, 128, 139–141, 146–150, 152, 156–157 © Alano Design/Shutterstock.com

THE ICE RAID

On a hill above Bardufoss airfield the Norwegian anti-aircraft gunners stood by. Their Bofors guns were half concealed in snow pits. A few yards away rose the pallid dome of the radar which fed information to the guns, tracking approaching aircraft. Inside the warmth of the radar cabin, the crew monitored a steady stream of approaching aircraft. The ACE Mobile Force exercise had begun and C-141s, huge long-winged jets, were bringing troops from the United States and Canada; stubby Hercules turboprops had already offloaded a British battalion from Tidworth in Wiltshire; the force commander had flown from his headquarters in south Germany yesterday; more troops were arriving all the time.

Also by Richard Cox

KENYATTA'S COUNTRY
OPERATION SEALION
SAM 7
AUCTION
THE TIME IT TAKES
THE KGB DIRECTIVE

THE ICE RAID

Richard Cox

ARROW BOOKS

A disaster always results in someone being promoted
T.X.H. Pantcheff, CMG

The line between courage and stupidity is sometimes rather thin
Major General Frederik Bull-Hansen

Arrow Books Limited
17-21 Conway Street, London W1P 6JD

An imprint of the Hutchinson Publishing Group

London Melbourne Sydney Auckland
Johannesburg and agencies throughout
the world

First published in Great Britain 1983
by Hutchinson & Co. Ltd
Arrow edition 1984

Printed and bound in Great Britain by
Anchor Brendon Limited, Tiptree, Essex

ISBN 0 09 937760 8

Acknowledgements

My thanks are due to numerous American, British and Norwegian military men for advice and assistance, especially General Sir Edwin Bramall, General Sir Anthony Farrar-Hockley, Major General Frederik Bull-Hansen, Brigadier Rory Walker, Colonel Mac Radcliffe, USMC, Colonel Myron Harrington, USMC, Lieutenant Colonel Peter Walter, the former Sysselmann of Svalbard, Mr Jan Grøndahl, and Mr Stein Sire of the TFDS shipping line. I must also acknowledge the insights into Marine Corps thinking given by Charles R. Anderson's vivid book, *The Grunts* (published by Presidio Press, San Rafael, California). Finally I thank Mrs Sheila Pincus and Miss Tina Grainger for their unstinting help in typing and revising the manuscript.

Richard Cox
Alderney and Nairobi

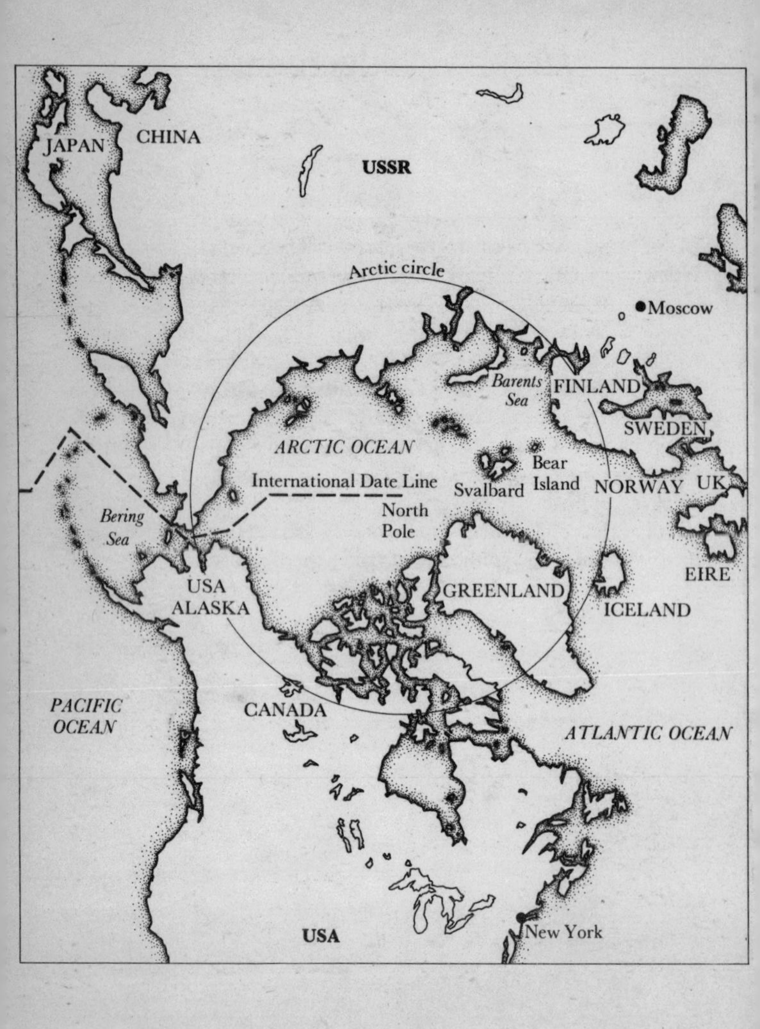
JAPAN
CHINA
USSR
Arctic circle
Moscow
Barents Sea
FINLAND
SWEDEN
ARCTIC OCEAN
Bear Island
International Date Line
Svalbard
NORWAY
UK
Bering Sea
North Pole
EIRE
USA ALASKA
GREENLAND
ICELAND
PACIFIC OCEAN
CANADA
ATLANTIC OCEAN
New York
USA

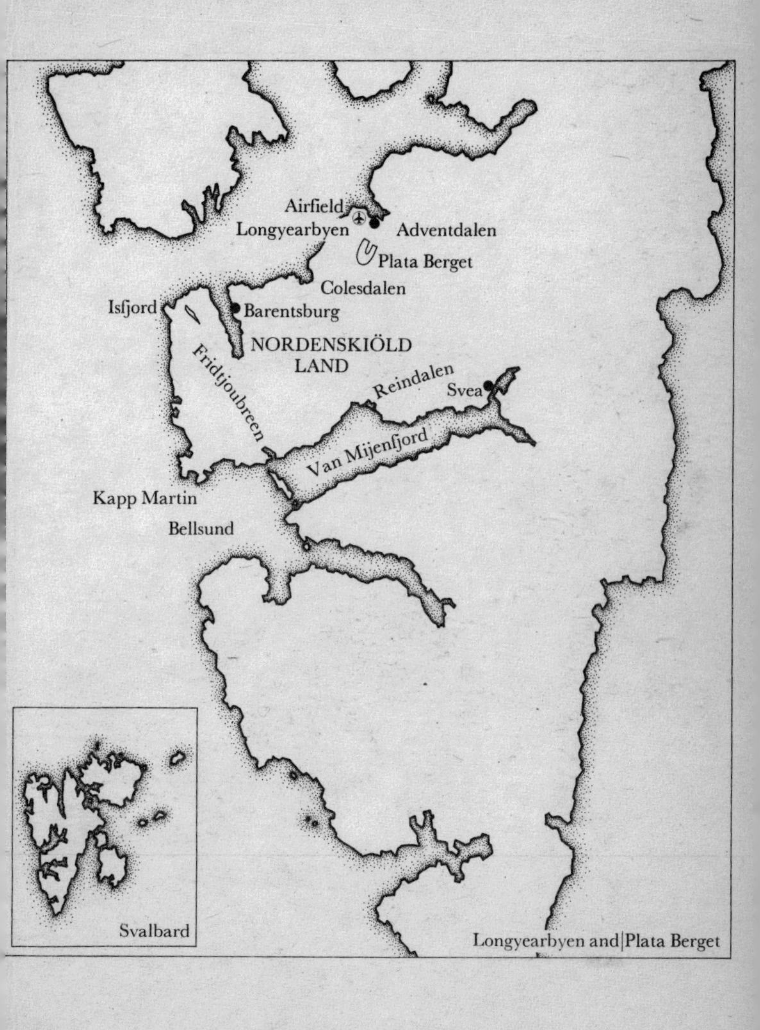

Longyearbyen and Plata Berget

1

It was early February and for nearly four months the Arctic landscape had seen no sun. The darkness was unending, although on cloudless days the snow reflected back a strange luminosity from the sky, creating an illusion of moonlight.

Standing at the control-tower window, Makarov accustomed his eyes to this twilight world. To the east he could discern the coal-loading jetty, projecting like a spike into the snow-covered ice, and the black shapes of buildings on the shore. To the west the broad reach of the fjord was a vast, white plain. It would remain frozen until early June. Directly in front of him lay the runway of Longyearbyen airfield, with a great mountain rearing up behind. Everywhere he looked there were mountains, the cruel peaks which had given this land of Spitsbergen its name. But the one dominating the airfield was different. It was flat-topped, the almost level crest a ghostly outline above which shone the pinprick lights of a few stars. Because of its flatness, this mountain was called Plata Berget.

In summer, so the Norwegian air-traffic controller had told him, the Plata Berget changed. In July the snow melted and a small valley cutting into the mountainside became alive with clusters of tiny red and yellow flowers, wild reindeer wandered the lower slopes, unafraid of man, and occasional polar bears came down from the interior to scavenge among the wooden holiday cottages on the shore.

But in February the Plata Berget was an icy, unyielding bastion and Makarov eyed it with the measured appraisal of a boxer weighing up an opponent. If this afternoon's events went smoothly he would soon be challenging the Plata Berget's isolation.

'Will our flight be on time, Mr Folvik?' he asked in thickly accented Norwegian. There was a staccato rhythm about Scandinavian languages which many Russians found hard to master.

'So far as I know,' the controller replied cautiously. He was a balding round-faced man whose amiable expression was heightened by spectacles. Folvik had seldom met the Russian and found his politeness unconvincing, worrying even. Makarov was thickset, of medium height, with a strong, broad face, his skin weathered and crinkled around the eyes, his mouth unyielding. His was a face carved from the hard timber of experience, the expression of a man accustomed to giving orders, not asking favours. Yet ever since his arrival to take over as local Aeroflot manager in October Makarov had displayed this same punctiliousness and Folvik mistrusted it. Just now, for instance, when he had heard Makarov's measured tread ascending the stairs, he had wondered if the Russian would take his boots off before entering the room. Norwegians were careful not to bring snow or slush indoors. Folvik was always half afraid that Makarov would neglect this courtesy and awkwardness would follow.

Accordingly Folvik walked across and made a show of checking the teleprinter messages. 'The flight departed Murmansk on schedule,' he confirmed. 'It's still due at 14.05. Do you expect many passengers?'

It was an unnecessary question. The airliner was obviously going to be full. An hour ago all five of the Russian helicopters based down the fjord at Barentsburg had clattered in, big-bellied machines originally designed for military use. They stood on the apron below, the yellow airfield lights casting long shadows from their rotor blades, making him curious. It crossed his mind that when the Russians flew in replacement workers they might deliberately include a number of good footballers. They were fanatic about winning every kind of sport against the West.

Folvik's hobby was running the local Longyearbyen football team. Only two days ago he had been present when the Norwegian Governor, the Sysselmann, had held a meeting with the Russian Consul from Barentsburg to discuss the summer's sporting and cultural exchanges between the two communities. They had agreed on a match in June, followed by a return in August. Folvik was determined that his team should win at least one. They would start indoor training this very week. Why the hell should the Russians always win, even if they did have twice the population to draw on? Why should their miners be better footballers than the Norwegians, when their coal production per head was so much lower?

The Russians' motivation often perplexed him. He felt that if he had been to the Soviet Union he might understand them better. For example, both Norwegians and Russians had the same gut problems in Spitsbergen, the only inhabited part of the Svalbard archipelago, an area which for centuries had been a no-man's-land on the edge of the polar pack ice, visited only by hunters, trappers and whaling expeditions. They were all here because an American called John Munro Longyear had pioneered mining coal in 1905. Munro had persisted despite the bone-chilling climate, despite earth so fiercely frozen that it threw up anything you buried, whether corpses or sewage pipes. Nothing ran underground in Longyear City, as he had optimistically named the settlement, except the mine shafts. However, the Great War had done what the permafrost could not: forced the American out. A Norwegian company took over his concessions and a Russian company took over the Swedish ones.

So far as survival and extracting coal went, Russians and Norwegians were in the same boat, except that since 1925

Svalbard had been ruled by Norway. But the Svalbard Treaty of 1920 gave all signatory nations, including the Soviet Union, equal right of access, and in practice the Russians were allowed to run their two company towns of Barentsburg and Pyramiden much as they wished.

So why, at the meeting on Monday, had the elderly and normally benign Soviet Consul wasted their time with a denunciation of the United States fleet operating in the sea areas between Greenland and Svalbard? Everyone present knew that the Treaty forbade the use of Svalbard for any 'warlike purpose'. The Sysselmann had reminded the Consul of this, but it hadn't stopped the old man. He'd wasted half an hour trying to blame Norway for the American fleet's actions. Makarov had been present at the meeting too and Folvik toyed with raising the subject now.

'It's unusual for your flight to be full,' he commented instead, losing his nerve.

'A visiting delegation is on board,' remarked Makarov shortly. He was under no obligation to reveal anything. Soviet citizens could come and go freely, without visas, as could Britons or Americans, Japanese or Swedes, or any of the others. He eyed Folvik, thinking to himself that here was a basically weak man, indecisive except within the parameters of his profession. He had advised Moscow that there would be no trouble on the airfield.

'Ah.' Folvik speculated on the delegation's purpose. The Russians had sent many scientific expeditions to Svalbard, though this was early in the year. 'Won't see much, will they?'

'They come to study winter conditions in the community,' said Makarov with impassive finality.

'Thank God the end of this winter won't be long.' On 14 February the thousand miners and officials in Longyearbyen, their wives and families, would feel a new stirring in their blood. On that day the earth's slow annual rotation around the sun would bring latitude 78 degrees N out of shadow and around noon the year's first sunrise would lighten the southern horizon. From then on the transition to the situation of constant daylight would be astonishing. By 20 April the sun would be shining at midnight.

'We're playing your boys at football in June,' Folvik said. 'This time we'll win.'

Makarov's blank expression reminded him that the Russian was a newcomer and as if to apologize he offered him coffee. 'It's only instant, I'm afraid.'

Makarov watched as the Norwegian moved to the back of the room, where cups and saucers stood on a cupboard and a small aluminium kettle was perched on an electric ring, incongruously old-fashioned compared to the sophisticated radio equipment under the windows. No, he thought, Folvik won't be a problem. But the Governor might, though any resistance he could put up with four policemen would be laughable.

Folvik made the coffee, stirring in powder from a jar with his initials marked on the lid to distinguish it from the other controller's supply. His christian name was Frederik, the result of his mother's imagining that her son might pursue a career in the arts. Frederik Folvik. She had visualized it on posters, in lights even, and overruled his father's protestations that such artificially alliterative names were uncharacteristic of Norway. 'Let the Swedes go in for that sort of thing,' he had said, scornfully. But she had prevailed. So now Folvik was known half-jokingly to his friends as 'FF', as if he were a business tycoon instead of a slightly diffident air-traffic controller.

The initials on the jars amused Makarov. How bourgeois men could become over private property! He accepted the cup of coffee readily, however. The desolate cold outside gave an added quality to hot drinks; one instinctively welcomed warmth of any kind in a climate where mere survival was a problem.

'So how do you like Longyearbyen?' asked Folvik.

For a second Makarov suspected the controller was reading his thoughts. 'They have given me good accommodation,' he said gruffly. The five members of the Aeroflot control team all lived in housing provided by the Norwegian government. 'And you?' he asked. 'Why do you stay here?'

'To save money.'

'Ah.' He was surprised by Folvik's directness. 'You get extra pay?' Makarov was receiving both hardship money and position pay for special responsibilities.

'No. Just low taxes. That's the main attraction of Svalbard – '

The tower radio interrupted as the pilot of the airliner far away began a transmission. 'Longyear Information, this is Aeroflot 207.'

Folvik put down his cup and moved to the centre of the long console to pick up a microphone. 'Aeroflot 207, this is Longyear, go ahead.'

Using the standardized English of civil aviation, the Soviet pilot announced his estimated arrival time and Folvik gave him the wind speed, ground temperature and visibility. The runway ran almost exactly east to west and with a westerly wind the pilot ought logically to select the direction into wind. But this involved descending into a long valley between mountains. The eastern approach never worried Scandinavian pilots: the airfield had a first-class instrument-landing system. But the Aeroflot captains usually avoided it, preferring the easy approach over the wide fjord, even though they would be landing with the wind behind them and touching down at a much higher speed.

'Landing will be runway one zero,' came the clipped voice of the pilot.

Folvik turned to Makarov. 'With a twenty-knot wind he should use the other direction. He'll have a very fast touchdown.'

'You cannot order him.'

'True.' Folvik was not a first-grade controller, only a secondary one. He could give advice, not commands.

'And you have no radar.' Makarov pressed the point

harshly. 'The pilot has to judge his own position with the landing system. He is at risk. Not you.'

Folvik merely shrugged his shoulders, his shapeless pullover seeming to emphasize his disregard. Why argue? If the Aeroflot pilot cared to risk running off the end of the runway, that was Aeroflot's affair.

'Commencing descent.' There was an edge in the pilot's voice too, a tautness at variance with the normal.

'Descend at your discretion. We have no other known traffic.' The phrases did indeed put all the onus on the pilot. Folvik noted the time on a routine slip of paper. He speculated on how important a delegation this must be. Presumably the Sysselmann knew about it.

A few minutes later they heard the jet pass over the airfield's radio beacon and Makarov strode to the window to watch its lights retreat down the fjord as it lost more height. At 3000 feet it would swing round to begin the final approach. He was openly impatient now, anxious for the action to start. He glanced down at the tarmac. A solitary Norwegian policeman, his blue uniform trousers tucked into long boots, patrolled near the helicopters, then disappeared from view into the large arched hangar which served as a terminal building. The policeman was unarmed. Makarov turned his attention to Folvik, who had sat down, microphone in hand, and was staring out at the dim white expanse of the fjord. If they follow my suggestions, Makarov thought, nothing can go wrong, unless the pilot misjudges the landing. He didn't believe this was possible, though the controller's warning troubled him.

The airliner was on the way back now, its navigation lights bright in the gloom like moving stars. Suddenly the landing lights flicked on as well, two dazzling eyes in the darkness.

The pilot's voice came over loud. 'Aeroflot 207 finals.'

'Runway is free for landing,' Folvik replied in the not quite idiomatic English he used.

Makarov stared out, the plane's lights blinding him until

it was about to land. When the beams swept only the runway ahead he could discern the shape of a three-engined Tupolev. God, it was going fast, too, touching down late and shooting past the tower. The engines howled as the pilot used reversed thrust to slow down. Snow whirled up behind. Out of the corner of his eye Makarov saw Folvik reach for the crash button which would send a fire crew racing out. That, more than anything, suddenly undermined his confidence. Involuntarily he screwed his fingernails into his palms, watching helplessly as disaster seemed inevitable. The windows were shaking from the roar of the engines and he could see the plane shuddering as the pilot braked. But it rolled on inexorably towards the end of the runway and the waste ground sloping down to the fjord.

'Stop, damn you!' Makarov half-shouted, agonized at this unanticipated threat to the plans he had so carefully consolidated.

Then, in what seemed the last yard of space, the Tupolev stopped and the cacophony of noise died away as it turned slowly and began taxiing back along the runway.

'That was close,' commented Folvik.

'I do not agree!' Makarov recovered his composure instantly. 'The landing was without fault. We have most experienced pilots.'

'Aren't you going down?' asked Folvik, as the plane came to rest in front of the tower.

Makarov did not reply. On the tarmac airport employees wheeled out steps. Four men walked to the foot of the steps, two of them Norwegian police, one the airport commandant, the fourth Makarov's assistant. The airliner's door edged open a crack, then was swung aside. A stewardess appeared in the doorway, called down to Makarov's assistant, and the first passengers emerged, all burly men in thick coats. Makarov half nodded in satisfaction. This was what he had advised: to maintain the appearance of normality to the last moment. Cut off from hearing voices by the windows, he had the momentary illusion of watching a silent film.

When a dozen passengers had descended and grouped themselves round the three Norwegians, the action started. A Soviet officer of the Border Police Troops, in uniform, appeared in the door and hurried down, followed by more uniformed police. They wore leather belts and holsters over their military greatcoats. The airport commandant stepped forward, protest evident in his gestures, but was firmly held back.

'What's happening?' cried Folvik. 'What are they doing?'

The Border Troops were streaming out now, jogging down the steps with ordered haste. The two Norwegian police struggled, but were pinioned by the original 'passengers'.

'You're not allowed soldiers here.' Folvik turned on Makarov, a mixture of indignation and fear on his face. 'I shall call the Sysselmann.' There was a direct line to the Governor's office.

'Please do so,' Makarov answered coolly. 'Tell him to come out here at once, because the Soviet Minister, General Boris Ivanovich Stolypin, will arrive in half an hour.'

'What do you mean?' Folvik was trembling. Through the windows he could see Russians running to take up positions round the building. They carried rifles and were obviously acting to a prearranged plan. 'What's going on?'

'Call Governor Prebensen. That's an order.'

Folvik picked up the telephone but still hesitated. 'What shall I tell him?'

Makarov suppressed a desire to pick up this woolly-headed idiot and shake him. Norwegians were to be treated with respect, those were his orders, and he could afford to be magnanimous now that the airfield had been seized without bloodshed.

'Tell the Governor a joint administration is being established which will reflect the political realities of Svalbard.'

'We don't want you here!' Folvik suddenly straightened up and faced the Russian. 'This is our country.'

The challenge was unexpected, but not strong enough to

catch Makarov off balance. 'You are wrong, my friend. We are the majority. There are 2300 of our miners. Only 1100 Norwegians. Don't you believe in democracy? The Treaty is out of date. Now, will you call the Governor or shall I?'

Reluctantly Folvik obeyed.

'The General's aircraft will be given absolute priority in landing,' Makarov added. 'And remember, so long as you cooperate with us, we shall cooperate with you.'

He left Folvik and went down to be ready for the Sysselmann's arrival, confident that the *coup de main* was irreversible. Sysselman Prebensen would protest but be forced to submit. They had been totally correct when they briefed him for this mission in Moscow five months ago. 'The Western European capitalist nations are divided,' Stolypin had told him. 'Their warmongering so-called Alliance is weak and at loggerheads with the Americans, who are giving us unbearable provocation in the eastern Atlantic. We must respond or they will make further aggressive moves. Svalbard can split the imperialist Alliance down the centre. You, Comrade Colonel, are privileged to serve the defence of our motherland in this hour of decision.'

'Well,' Makarov said to himself as he emerged on to the windswept tarmac and glanced up at the outline of the Plata Berget. 'The imperialists will have to think twice about what to do when I get the radar up there.' He shared the Politburo's view that no Western leader would risk war for this chunk of Arctic wilderness.

2

The barmen wore jeans, cowboy shirts and Stetson hats, imitation Colts hung in holsters from their belts. They were serving in a reception room at the Brussels Hilton and were dressed like this because drinks were on the State of Georgia tonight. They only stopped pouring scotches and bourbons for the business guests when a Georgia representative announced a presentation.

On a low stage at the end of the room an American girl dressed as Scarlett O'Hara, dark and extremely pretty in a flouncing white crinoline skirt, faced the audience. She glanced as if for reassurance at the frock-coated figure of Rhett Butler beside her.

'The guy looks even more stupid than we do,' whispered one barman to the other. 'Wouldn't mind the doll, though, would you, Erich?'

'Not tonight.' Erich Braun cut his colleague short. He felt a fool in the cowboy gear too, but he was watching the crowd and didn't want to be interrupted. He learned a lot from watching and listening in this hotel. Even at an American trade promotion one could learn.

On the stage Scarlett O'Hara dimpled and curtsied as the guest of honour stepped up. He was the American Ambassador, a heavily built man in a blue suit.

'We hope you will accept this small memento for your library, sir.' She smiled, holding out a book.

The Ambassador took the volume, hesitated, then kissed her cheek. 'You have given me a marvellous welcome,' he said. 'One I shall never forget.' The audience applauded and he continued his thanks: 'Honoured to be present . . . close ties between the United States and Belgium . . . warm affection . . . one of my predecessors in this post came from Georgia. . . .'

Erich studied the audience, smiling to himself as a fat Belgian woman in a silk dress slid out her hand for the last slice of strawberry flan on one of the tables. Greed, he thought, when they're older it's even more compelling than sex.

At length the Ambassador's speech finished, the 'Gone with the Wind' couple swung into a self-conscious waltz and the mass of guests turned thankfully back to the bar tables with a restrained clamour of orders. More food arrived, cuts of rare beef with jacket potatoes, melon and peaches, skewered ham, blueberry cheesecake, prawns, pecans in chocolate. Georgia's Department of Industry and Trade was not stinting its hospitality. Erich realized he was going to be late. Mechanically pouring more bourbon for a diplomat, reaching for ice cubes, he reckoned he could count on Sandra waiting. Hell, he thought, of course she'll wait. She wants it like the fat one wanted strawberries. What am I getting nervous about? None the less he checked his watch repeatedly. Tonight he would be pushing his luck. He didn't want her uptight for any reason.

The manager of the Duc d'Arenberg bowed welcomingly as Sandra Singleton entered, noting that the time was precisely eight o'clock. She was his most punctual patron, but his eyebrows lifted a fraction as he saw she was alone. A woman of sense, he felt, should always be a little late. This English girl was doing herself no service, especially with the particular man she had in tow. As he ushered her to the table the couple always had in a corner of the restaurant he speculated further on this theme. Madame Singleton must be nearing one of the most susceptible ages for a certain kind of woman, the time when child-bearing is within measurable distance of being difficult, when the fulfilment of marriage would sour from a romantic dream into a psychological necessity. He knew perfectly well that she was unmarried, yet called her madame from traditional Gallic politeness.

'*Voilà*, madame.' He gently shifted the table as she sat down. 'Your usual drink?'

'Thank you.' She smiled up at him. She really liked the Duc d'Arenberg. It was intimate, full of atmosphere, attentively run, just the sort of place you would hope to find in a quiet street near the Royal Palace. It was also ten minutes' walk from the Hilton, which made it convenient for Erich, though it was the other side of town from the Square Marguerite where she lived. She called the restaurant 'our place' with a loving intonation which she had once noticed made her boyfriend's smile tighten. 'But it is, darling,' she had protested and he had laughed quickly.

The manager returned with a bottle of the Muscadet she liked. Though it made him wince to add blackcurrant liqueur and ice to a decent wine, he was better at dissembling than Erich. Now he mixed the kir and set the glass in front of her with a flourish. Then he handed her the extensive menu. 'I can recommend the sole *bonne femme*. Our fish is excellent today.'

'I won't order until my friend arrives.'

'Of course.' The manager withdrew, positioned himself where he could study her briefly. She was definitely past thirty, he decided. What did they call that complexion? The English rose. Well, this one was losing its bloom. True, her features had a pleasant regularity, but only the freshness of youth could have made them seem pretty. Nor did she have obvious sex appeal. So what attracted that undeniably good-looking, undeniably younger German to her? Bed, maybe. In the dark all cats are black. Or could she be paying him? No! Such an idea was idiotic. But just what was this well-spoken middle-class woman, who he knew worked at the NATO headquarters, doing with a barman from the Hilton? By accident the manager had discovered where his patron was employed and it had surprised him considerably. He had imagined the man was a journalist. His speculations were interrupted by the arrival of Erich Braun himself, looking like a writer in a brown corduroy jacket and flowing silk tie.

'Good evening, sir.' The manager moved forward, though his welcome was less effusive than it had been for the lady. A barman was still a barman, even if he did have the money to eat at the Duc d'Arenberg. 'Madame is waiting.' He ushered Braun across the room.

'I am so sorry, darling. I couldn't get away.' Erich leaned across the table and kissed her hand with affection, then took the chair opposite and asked for a Cinzano and soda. He would have liked a whisky, but the Cinzano fitted his image better. In Belgium whisky represented conspicuous consumption, expense-account living. He always told Sandra that although he was happy to splurge on the restaurant, he had no expense account. There would be whisky enough back at her apartment, stocked with duty-free bottles from the NATO commissary.

'Don't worry! I haven't been here long.' She gazed at him tenderly and with approval. His thick dark hair curled down over his ears and framed a handsome, slightly sensual face. She felt herself tremble when he smiled back. 'Why did they want you this evening anyway?'

'A big reception. They needed extra staff.' He half snorted with indignation at the memory of the cowboy gear but did not mention it, afraid to make himself foolish in her eyes. 'There might have been a story for the magazine in it too.' He shook his head disgustedly. 'What can one make of a bunch of fat Belgians eating their heads off and listening to the American Ambassador's platitudes? Null, null, null.'

'Poor darling. It must be terribly difficult being a freelance.'

'If I can only make a name for myself! I know there'll be a feature writer's job going on *Der Spiegel* next year. All I need is a few exclusive stories. When a small magazine like ours beats the big boys at their own game they'll soon take notice.' He paused while a waiter deposited the tall glass of Cinzano soda. For once the interruption was welcome. He was going too fast, there was time in hand, his indignation could be built up later. The immediate thing was to give her an unforgettably enjoyable evening.

'Did the manager suggest anything special?' he asked solicitously.

'The sole.' She scanned the impressive menu, savouring the names of the dishes. All too often her supper was cooked by herself, for herself, alone in the apartment. Erich worked five evenings a week and at weekends if he wasn't too busy they usually went out to the flying club at Grimbergen where they had originally met. Going to the Duc d'Arenberg was a special treat, spiced by the memory that it was after a dinner here that they had first made love. 'Oh dear,' she sighed. 'My eyes are always bigger than my tummy.'

Although he did not understand the phrase, he nodded sagely. 'If you are in doubt, accept the sole.'

'The Muscadet goes well with it.'

'Then you have decided.' He glanced up at the waiter, hovering with order pad in hand. 'For madame, the sole. For myself, the *faux filet*.' He turned back to Sandra. 'You can taste some of mine, too. Just to be greedy.'

'You are a wicked man, Erich. Reading my thoughts.' She was delighted. He was so clever at pleasing her in small ways.

He reached out and touched her fingertips, looking into her eyes. 'I hope we are getting to know each other.'

The way he said it, in a low voice, made her blush momentarily. The words sounded halfway to a proposal, a thrilling confirmation that their relationship, only three months old, really was deepening all the time.

'I'm so happy, Erich,' she said simply.

He smiled at her again, his gaze wandering across her

face, clinically noting the lightly made-up lips, the perceptibly receding chin, the wide-set hazel eyes, their colour carefully accentuated by greeny shading on the eyelids. Erich did not often look directly at her. His declarations of love were normally made to a point a few inches above her head, or while he stroked her hair, a position which made gazing into her eyes conveniently impossible. Those eyes, he decided now, were definitely her best feature. Even so they reminded him more of a gazelle than a human; limpid, wide and somehow frightened. However, there must be tenacity of purpose behind them or she would never hold down her present job, let alone have learned to handle a light aircraft successfully. But he wished she would keep off those appalling semi-provocative schoolgirl phrases into which she lapsed whenever they were alone.

Inevitably he let slip a hint of his ruminations. 'Did you ever think of becoming a blonde? It would suit you.'

'Really?' She blushed again. She had indeed considered so doing, but never plucked up courage. 'What would everyone at work say?'

'Doesn't General Anderson like blondes? I thought all Americans did.'

'I wouldn't do it to please my boss, silly. I'd do it to please you.'

'Then why not?' For once his enthusiasm was unforced. Since he had to continue courting this woman, he would much prefer her a blonde.

'I will!' She was pleased and excited. 'You make me feel so alive, Erich. I'm a different person with you.'

'No, my love. It is I who am lucky.'

Again the waiter interrupted conveniently, saving him from further false sentiment, and when the first course had been served it was easy to switch the conversation nearer his interest. 'How are things at NATO anyway? Busy?'

'It was a bit hectic today,' she admitted.

'Is it true that the Soviets refuse to hand back Svalbard? It came through on Reuters and I could hardly believe it.'

Sandra was curious. 'What did Reuters say, then?'

'They carried a long Norwegian government statement. About legal sovereignty. But one of the others took it. That's the problem of sharing an office.'

Sandra nodded sympathetically. She knew all he could afford was a desk in the corner of a room used by several other correspondents, all full-time journalists based in Brussels to cover NATO, the Common Market and the European Parliament. As a freelance on a retainer for a small German weekly, Erich enjoyed no priority among his fellows.

'Would it interest your editor much anyway?' she asked. 'Surely the Arctic is a bit remote for people in Munich?'

'Not if the story's strong enough. And this one could turn into a major confrontation between the Soviets and NATO.' He spoke with assurance, deliberately begging the question. Sandra's boss, General Hiram G. Anderson, was chairman of the NATO Military Committee, which advised the NATO Council. Every aspect of the West's defences was known to the Military Committee, though not necessarily in detail.

She hesitated, thinking of the worried activity in her office today. General Anderson had spent a considerable time on the secure telephone line to the Supreme Allied Commander down near Mons and when she had to go to the Situation Room, the most secret part of the Military Committee's offices, two colonels had been busy with maps of the Svalbard archipelago, the Barents Sea and northern Norway. Not that there wasn't always keen interest in the extreme north. The largest concentration of military power in the world was based on the Kola Peninsula, on the frontier of Norway, and that military power was Russian. But what could she tell Erich? She hadn't even known where Svalbard was until yesterday and didn't yet understand the issues involved. Nor was she supposed to talk about her work. But it couldn't do any harm to admit that the news continued to cause concern. 'Well,' she conceded, 'it has set the cat among the pigeons, all right.'

'The cat . . . ah.' He appreciated the cliché, saw a chance and responded provocatively. 'I thought they were hawks at NATO, not pigeons!'

'Really, Erich!' She tittered in spite of herself. 'Well, they're certainly flapping round in circles.'

'Svalbard catches them by surprise, yes?'

'Did it!' She giggled. In NATO's multi-national secretariat a colonel was practically the lowest form of military life while generals were two a penny, and there had been something very comic about so many senior officers all demanding information at once in a variety of languages. 'I shouldn't laugh.' She giggled again. 'But it does have its funny side. They are so very serious, Erich.'

'I would say it was a very serious subject. So what happens next?'

'You'd better ask the Russians that.'

'Don't you understand?' He did not like her tart riposte. 'Svalbard is Norwegian territory, therefore it is NATO territory. Can the Western Alliance allow this to happen?' He leaned across the table slightly, trying to impress her. 'Here is the biggest story to come out of Brussels for years. Couldn't you help me with it, give me some guidance? The magazine is certain to want coverage of how this crisis develops. It could be the chance I've been waiting for.'

'Darling, I really don't know anything about it. Except that this Svalbard place is demilitarized, so it must be a political business, not a military one anyway.' She was troubled now. She would have so liked to help him, she knew how desperate he was to make a success of his journalism, to escape from the need to have this degrading part-time job as a barman. 'Can't the NATO Press Office give you a briefing?'

'Those . . . !' He swore. 'They are making no comment. Null, null, null!'

She picked up her knife and fork determinedly. 'This

lovely fish is getting cold. Don't let's spoil the evening by arguing.' She tasted a little. 'It is delicious.'

Reluctantly he began to eat too. Seeing the resentment linger in his eyes, she said remorsefully, 'Honestly, Erich, I don't know anything about what's going on.'

'Not at this moment perhaps, but you will do.' He knew she would because the Military Committee had held its weekly meeting today and she would type the account of the proceedings.

'Well, tomorrow is another day, isn't it?'

Suddenly he realized he was pressing too hard and changed the subject to the redecoration of her apartment.

Her relief was obvious. 'Didn't I tell you? The new wall-paper is up. I can't wait to show you. I'm sure you'll approve.'

An hour later they were on their way there in her Mini Metro, she happy on the wine, he tense and sober. Braun was a realist and kept his consumption of alcohol down on his nights out with Sandra. Even a physically fit twenty-eight-year-old's sexual prowess could be affected by drink.

The Square Marguerite was a pleasant place to live. The trees in its centre, albeit now bare and wet in the February rain, gave it distinction and the recently built apartment block was luxurious, with marble floors in the lobby. Sandra herself was always the first to admit that even as a senior secretary on the highest levels of pay available she could never contemplate such luxury in London. At the best she would have been able to buy a lease in Battersea or Fulham and would probably have had to share with a friend in order to pay the mortgage. But with NATO's minimal tax rates and generous allowances, she could afford the Brussels equivalent of Mayfair, and have the place to herself. What she had at first been less willing to admit was the amount of time she did spend there alone, half-heartedly watching the TV. Eventually, to get herself into a new social circle, she joined the flying club out at Grimbergen. The flying had succeeded beyond all expectation. It had introduced her to Erich and Erich, as she soon discovered, was an accomplished lover.

'So what do you think of it, darling?' She proudly showed off the new decorations of her living room. 'All done in three days.'

'Very nice.' There was genuine warmth, even a touch of envy in Erich's response. The room was large, with picture windows overlooking the square. She had chosen a bamboo-patterned wallpaper that unquestionably enhanced the long pale beige curtains, though it made the sofa look dowdy. 'All you need now is a photographer from *House and Garden*.'

Sandra flushed with pleasure, then put her arms round his neck and kissed him. 'I'm so glad you like it. Now, we shall have a drink to celebrate.' She moved across to a two-tiered drinks trolley, fashioned of inlaid rosewood with large brass wheels, and picked out a bottle of Southern Comfort. 'Your usual, sir?'

'A small one, barman, please.' He was getting better at matching her mood.

She poured him a measure in one of the little Irish crystal glasses which stood on the trolley's top gallery, gave herself a crème de menthe and turned on the cassette player. A Cole Porter tune filled the room.

'I get no kick from champagne,' ran the song.

She danced a step, almost upset her liqueur, then settled on the sofa. She was slightly drunk and very content.

Erich switched off one of the table lamps, then sat beside her and squeezed her hand while he sipped his whisky.

'I forgot to say how beautiful you are looking tonight,' he said softly. 'That was bad of me.'

'Thank you, darling.' She turned and kissed him on the cheek.

Time to start, he decided, put down his glass carefully, and kissed her on the lips, at the same time reaching to caress her breasts. She responded avidly, running her tongue inside his mouth, then moaning with anticipation as he undid her dress. Already, in these short months, the way they began had become a ritual. But this time, instead of guiding her to the bedroom before her excitement became too intense, he rolled her off the sofa on to the thick white carpet and took her then and there, driving into her as if possessed until she was gasping with pleasure. He reckoned, correctly, that she had never been made love to like that before.

'God, I want you so badly,' he murmured. 'Oh, I want you.' No sooner was it over than he lifted her up, her arms round his neck, and carried her through to the wide bed which she had bought soon after their affair developed. There, pausing only to drape her clothes over a chair and strip off his own shirt, he flung himself on her again, muttering affirmations of love and passion.

'Oh, darling, I love you,' she said, almost crying.

Eventually they lay together, her head cradled in his arms, and he began to talk quietly about himself and his ambitions, about the false start in commerce which he had hated, about the unexpected chance to write for the magazine in Munich and all that meant to him, though he still had somehow to send money to his widowed mother in Berlin. If only now he could make an outstanding series of reports on the East-West crisis which must develop over Svalbard, then he would be sure to get the job on *Der Spiegel*. And as *Der Spiegel*'s offices were in Berlin, he would be able to see his mother much more often and Sandra would find Berlin a wonderful place to live, if only she would join him there. So couldn't she somehow help him with a little insight into NATO's thinking over Svalbard?

Half asleep, exhausted, Sandra none the less preserved enough sense to say that important decisions were often made on the telephone and she didn't necessarily even know about them.

So he talked more about their future together and how he knew his true vocation was as an investigative writer. Eventually, as he began caressing her again, reawakening her sexuality gently, she consented to help him if she could.

With delights and fears threshing through Sandra's mind, one thought began to predominate: she would not be able to bear losing Erich. He was all she had ever dreamed of. Virile, exciting, ambitious, loving: the ideal antidote to the artificial life at NATO headquarters, with its fleeting liaisons and barbed gossip. Julia might have fun with her captains and colonels, but in their home countries they all belonged to other women. Erich had no absent wife waiting elsewhere. He was all hers. So she would try to help him.

Lying content in his arms, she reflected silently on what to do. There was no question of giving anything vital away, nothing secret, certainly not. She was well enough schooled in the disciplines of security to appreciate that. But all he wanted was the guidance he could not get from the Press Office. She could safely hint at things without stating facts. She remembered those childhood treasure hunt games. 'You're getting warmer . . . much warmer . . . no, it isn't locked up . . . getting colder now . . . yes, yes, warmer again.' Excitement mounting until the chocolates were revealed behind a flower vase. That was the way she would guide him, she decided, and she would do it because she so desperately needed him and now, wonder of wonders, he needed her too. She knew she was quite astute enough not to give away any real military secrets. So she kissed him sleepily and said how much she would like to meet his mother and he replied that he hoped she would soon.

But for his part he wondered if the complicated arrangements would be worthwhile. What he had not told her was that although he travelled on a West German passport, his mother lived in the East.

The Stars and Stripes hung limply from one polished flagpole behind the wide desk, while the blue NATO flag with its four-pointed star dangled from another. There was a conference table and chairs. Around the walls hung framed photographs of United States warships and a number of presentation plaques commemorating visits to foreign navies.

Overall the office of the Supreme Commander Allied Command Atlantic was workmanlike rather than opulent, with few reminders save a wall map that the Admiral co-ordinated the North Atlantic Treaty nations' defences of an immense expanse of sea, roughly a third of the ocean area in the northern hemisphere, stretching from the North Pole south to the Tropic of Cancer and from the United States east to the coastlines of Europe and North Africa. The office was in a red brick building at Norfolk, Virginia, one of the oldest and most famous bases the US Navy possessed.

From here SACLANT, as the Admiral's title was abbreviated, also acted as the Commander in Chief of the United States Atlantic Command. 'Wearing two hats', as servicemen called the dual responsibility, made him one of the key officers in the Atlantic Alliance. As SACLANT he and the European C-in-C, known as SACEUR, were subordinate only to General Anderson and the Military Committee in Brussels. In the United States hierarchy he was answerable to the Joint Chiefs of Staff and ultimately to the President, who was the Commander in Chief of all the nation's armed forces. This morning John H. King, the Admiral holding these prestigious posts, was wearing his NATO 'hat' and had his British deputy in attendance.

Admiral King was in his mid-fifties, grey-haired and, unlike most of his contemporaries in commerce, fighting a winning battle against fat. He looked and was extremely energetic, driving his international staff hard. As in most large corporations, the donkey work was done by underlings who rarely met the great man. But their success could not be evaluated in terms of the bottom line of a balance sheet, as if they were rising business executives. Come the end of the year, the greatest achievement they could normally look back on was that by patience, diplomacy and determination they had helped hold together the Alliance for a further twelve months. To a fire-eater like the Admiral this was not always sufficient satisfaction. Particularly when there was a crisis and the NATO Council could not decide what to do.

On this February afternoon, when the snow lay thick on the Blue Ridge mountains and the weather down at Norfolk was as grey as a battleship, Admiral King was questioning a lieutenant colonel of the Marine Corps who had joined his staff as an expert in Special Operations, while the British Admiral Frank Shiplake listened.

'So you've been up in this Svalbard place, Peterson?' King demanded.

'In 1976, sir.'

'How did you fix that?'

'I requested leave, sir. Took the Scandinavian Airlines flight up. Longyearbyen airfield opened the year before.'

'Didn't you need authority?'

'No, sir. Any national of the Treaty signatories can go without a visa.'

'The hell they can! Soviets included?'

'Correct, sir. Had to take my own camping gear, though.'

The Admiral turned to the Briton. 'Do your goddamned officers go freebooting around areas of potential conflict?'

'I know an army captain who won a Croix de Guerre with the French Foreign Legion while he was on leave. He was reprimanded and forbidden to wear the medal.'

'Hear that, Peterson? You're lucky not to be court-martialled.' Admiral King smiled crookedly. 'Thank God someone shows initiative in this damned Alliance. As of now you're assigned to our Svalbard project, right?' He moved across to the wall map, as if looking at it helped him concentrate. 'We have a problem. What the hell are we going to do if they militarize that Longyearbyen airfield? They can hardly make a naval base in the ice, but they can put in early warning radar and hardened shelters for strike aircraft.'

'Construction would be pretty difficult in winter, sir. Even in midsummer the earth's frozen to 150 metres down.'

'That's the kind of thing I want to know, Peterson. We have a range of options, right through from doing nothing to sending in the Marines or threatening to bomb the bastards. I want the facts on which we can base decisions. You get over to Norway and dig them out, OK? We seem short on ground-level information here.'

King studied the Marine's expression for a moment, wondering how much intellectual acumen lay behind the rugged face. Superficially Peterson was a lean, tough-faced soldier with a nose which had been flattened in a brawl and a scar running down the left side of his chin. Yet for all the short-cut hair and weather-beaten complexion, the Admiral knew he couldn't be a typical leatherneck. One reason was his eyes. They were a penetrating pale blue. Not the eyes of a man who intended to spend the rest of his life saying 'Yes, sir' and 'No, sir'. The second reason was his record, which King had perused before this interview. Peterson's career had been unusually varied.

'They tell me you're pretty sold on Norway,' King remarked in a more relaxed voice. He could afford a few minutes more with this man. He had plans for him.

'I guess so, sir.' Instinctively Peterson kept his answer terse, then thought better of it. If the Admiral really wanted to know he'd find out anyway. 'I got to know the country pretty well when I was an Embassy guard.' He elaborated. 'Before I went to 'Nam.'

That hardly expressed his feelings either; what he had enjoyed most were those long challenging cross-country ski runs through the mountains and the warm conviviality round a fire on winter nights. He had made good friends during his year and a half in Norway. But those days had quickly become a forgotten life, thrust aside by the harsh experience of Vietnam. He had been commissioned in 1967, the year before the Tet offensive, when the whole ball game changed.

'Yes, sir,' he said, forcing his mind back to the subject, 'I got a lot of time for Norwegians.'

'How do you think they will react now?'

'I'm no kind of a politician, sir.' Peterson hesitated fractionally, then was characteristically blunt. 'I guess they'll protest and that's about all. They always did play things pretty soft where the Soviets were concerned.'

'They stood up to the Soviet Navy over that Hopen Island incident,' remarked the British Admiral sharply. 'I seem to recall a Norwegian corvette telling a Russian cruiser to get lost.' When a Soviet TU-16 bomber loaded with electronic surveillance equipment crashed on Hopen Island, just south of Svalbard, in 1978, the Norwegians had refused the Russians access to the wreckage until they had checked it themselves. 'The story I heard was that the captain of that corvette threatened to open fire. I wouldn't call that soft, Colonel.'

'I'm not criticizing the Norwegian military, sir. Only questioning their politicians.'

'Well, thank you, Peterson.' King had heard as much as he wanted. 'You better get packed and ready. One thing. The official reason you're going is our participation in the Polar Express exercise. Remember that.'

'Yes, sir. I sure will, sir.' Peterson grinned, snapped to attention and left the room.

'Like the boy who's promised the candy,' commented the British Admiral wryly.

'I can use enthusiasm. Let me tell you something, Frank. That officer is one of the best I've got and you know how his career began? He was busted for assault when he was seventeen and a South Carolina judge said he could either go to jail or join the army. He chose the Marine Corps. You ask him and he'd probably say the draft would have got him anyway, but I think there's more to it than that.' He was about to expand on this, then decided against it. There was no reason to reveal even to a close ally just how Peterson had become an expert in Special Operations. Or that he was convinced some kind of special operation would have to be mounted in Svalbard, whether NATO agreed or not. He reverted to the main subject. 'We just have to send more Marines on Polar Express,' he said. 'Show the flag more conspicuously in Norway next month. That's the minimum reaction I'll accept.'

'Show the NATO flag or the Stars and Stripes?' The Briton glanced at the two limp standards behind the desk. 'Suppose the Norwegian government is still worried about being provocative to the Soviets? Neutralism has gained a lot of ground in Scandinavia lately.'

'Do we counter a direct Soviet threat by sitting on our asses?' The American cleared his throat noisily, a sign of exasperation. 'Listen, Frank, this is a well-considered Soviet reaction to our putting cruise missiles aboard carriers in the eastern Atlantic. And to our operating in the Barents Sea where the Defence Secretary declared we have a strategic stake as long ago as '79.'

'Your Defence Secretary, not anyone else's.'

'Are you telling me NATO wouldn't back us up on this? Goddamn it, who's been protecting Europe all these years?' King drew a deep breath.

'Your President and your Defence Secretary are going to have to accept an unpleasant fact,' said the Briton quietly. 'You've made the Soviets lose face in what they regard as their home waters. Now they're thumbing their nose at you by taking over what is technically NATO territory, even though its demilitarized.'

'So NATO should react.' King was vehement. 'That's the President's view, I can tell you.'

'In my opinion the European nations will only protest. Svalbard is no threat to them.'

'It is to the United States. Look at the goddamn map, Frank. It puts the Soviets 800 miles closer, it gives a whole new dimension to their missile superiority.'

Consideration of the map underlined the point. The shortest route around the globe from the heartland of the Soviet Union to the heartland of the United States passed over the Arctic. Establishing early warning radar on Svalbard would give the Russians greatly improved ability to react to incoming missiles. If they also based interceptor aircraft at Longyearbyen, they would be able to attack and shoot down both bombers and low-flying cruise missiles far out in the Atlantic.

'I don't dispute any of your strategic conclusions,' the British Admiral insisted, 'but the Norwegians have already demanded an emergency meeting of the United Nations Security Council. They'll look for peaceful solutions at any price and they won't welcome American interference in case that precipitates a total Soviet takeover in Svalbard. I hope I'm wrong, but as I see it the Soviets have got the West very neatly divided.'

'So tomorrow I go to Washington and tell our Joint Chiefs of Staff we're faced with what could be the worst situation since the Cuban missile crisis and there's not a damn thing we can do?' King swung his gaze away from the wall map and faced his deputy. 'The hell with that. What I want is a summary of NATO's capabilities, starting with the logistic time scale for boosting this exercise. Have the international staff begin work on that, Frank. Right away.'

When his deputy had gone, Admiral King stood studying the map again, considering possible moves. Svalbard was indeed a long way from Europe, 600 miles farther north than the North Cape itself, 2000 miles from the NATO Council's building outside Brussels. Although, in NATO terms, it lay within the Atlantic Command, not the European, he couldn't initiate a move there himself. Svalbard was Norwegian territory. The Norwegians would have to request assistance first. But would they? Hell, he thought, I have the right of access to NATO heads of state. I'll damn well use it.

The British had surprised the world with their recapture of the Falklands from the Argentinians and Svalbard was a lot closer to Norway than the Falklands had been to Britain. Maybe the Norwegians could be persuaded to act. That Hopen Island incident suggested they were tougher than people thought.

As he pondered, ideas began to clarify in Admiral King's mind. True, NATO as a whole probably would decline to respond sufficiently. But Norway, the United States and Britain might do something jointly. If the Soviets did put radar into Svalbard, a joint operation could be mounted to take it out: a quick, clandestine raid to show the Soviets privately that there were limits to what the United States would accept. He began mentally revising his brief to Peterson. The Marine colonel would be a key man in any such special operation and the more he thought about it, the more he knew there was no time to be lost. The Russians must have been preparing their moves for many months. They would act fast now they had started.

Tom Peterson reached home late that evening. Home was an attractively designed modern house at Virginia Beach, out near the estuary where the Elizabeth River meets the wide James River and conveniently close to Norfolk. The area was pleasant, countrified and favoured by middle-rank officers, for whom government accommodation was in short supply. Anyhow, as Tom's wife Nancy observed, the way inflation was hitting America it was crazy not to buy something somewhere, even if they could be housed by the government. She was from Minnesota and, quite apart from her immediate family's thrifty farming traditions, she had a typically middle-American belief in the virtues of real estate.

In this way Nancy was more down-to-earth than her husband. As a bachelor he had been content with whatever quarters the military provided when he wasn't out on exercises or operations. They had scarcely married before he had been sent off again and his appointment to the staff at Norfolk came as a godsend to her. She had found him shockingly slow to learn about growing roots. The promotion helped, though. There was a huge psychological divide between the ranks of major and lieutenant colonel. As a colonel he had broken through to the senior ranks of the

Corps. So they had bought this house with its big kitchen and three spacious bedrooms and begun making their first real home together.

Now they needed that home more than ever. A few days ago she told Tom that she was late this month. Today she had called at the clinic on the way back from school and the test was positive. She was pregnant. She felt excited and frightened and deliriously happy, all in one swirl of emotion. So why, why did he have to be home late on this evening of all evenings? By the time he had slung his peaked hat on a peg in the hall and shrugged off his uniform coat she was more than impatient.

'You warned me you'd be late, Tom. But not this late.'

'I'm sorry, sweetheart.' He kissed her. 'All hell broke loose this afternoon.'

'It did at school too,' she remarked tartly. 'Didn't I tell you the ninth grade are known as the hell-raisers? And I've cooked a special dinner for you.'

'Sure smells good.' He sniffed appreciatively. 'Could that be chili?' It was his favourite dish.

'It could. And you'd better come and eat it just as soon as you're ready.' She relaxed a little, walked back towards the open-plan living room and caught sight of herself in her treasured antique mirror, instinctively raised a hand to tidy her dark curling hair.

'Don't worry, honey. You're as beautiful as ever.' He followed, put an arm round her shoulder and nuzzled her cheek.

Their joint reflections smiled back from the mirror, framed like a portrait photograph. Nancy and Tom, the happy couple in their second year of marriage.

A confusion of memories swept through her mind. She saw the image of herself, lying awkwardly in the snow at Aspen, her shoulder starting to throb with pain from her fall, and heard again the deep voice from above. 'Hey, are you OK?' She had looked up, gripping the ski stick, and tried to raise herself. Pain had shot through her arm so viciously she cried out. 'I don't think I am OK,' she had managed to say. Then through a blur she had watched this pleasantly ugly man in a bright red parka lean down, felt his arms lift her. It had taken him only moments to identify the dislocation. He carried her out of harm's way to rest against a fir tree, wrapped his parka around her trembling shoulders, told her not to move, and skied fast down the mountain for help. Soon after he had helped bind her on a stretcher and he had been waiting there at the hospital when she emerged, her arm in a sling. A few weeks later they had become engaged. How on earth would she have met a Marine otherwise, least of all this one who was so often away and who was so absurdly reticent about his last post in the Delta Force? Well, she thought, the Marine Corps has had first claim for long enough. I want him home when I have our baby. In fact she would have liked him to quit the Marines altogether and settle down for good.

'You look great,' he said, squeezing her affectionately. 'I'll be down again in seconds.' He vanished from the portrait, leaving her gazing only at herself, wondering. Had he forgotten she was going to the clinic? The images in the mirror were like their life. Now he's with me, now he isn't. God, she thought, the baby had better make a change, because I do not intend to be any kind of a typical Service wife or mother. When older friends asked why she went on teaching school she joked about avoiding the wives' coffee mornings. More truthfully, she felt she would be losing part of herself if she didn't continue working. The kids she taught and this house at Virginia Beach were the real world, a useful reminder that neither she nor Tom belonged exclusively to what Marines called the Green Machine.

By the time Tom had come down again, Nancy had decided to try him out by saying nothing about the test until he asked. The chili was a success. She never doubted it would be. The spicy peppered meat matched one side of her husband's character and she had managed to get a particular Texan mix called Two Alarm Chili, which her cookbook rightly said had 'the bite of hot lava'.

'Boy, this brings back memories.' He ate enthusiastically.

'You mean of your Mexican girlfriend?' she teased. 'You had some steamy señorita stashed away down there, I know it.'

Predictably, he was embarrassed. 'Sweetheart, you know I never talk about that trip, but I swear señoritas did not feature any place.'

'Since when did the CIA cease chasing women?' She suddenly felt buoyant and devilish, quite unlike her usual self. She had never troubled about his past before. 'Don't try to kid me, Tom. What were you up to there anyway?'

Mexico had merely been a staging post for an operation while he was on detached duty with the CIA. The final destination had been farther south. He tried to laugh off the subject. 'The guy who first cooked chili for me did it in a pan you wouldn't have inside the house and the meat, well, I never did exactly discover what animal the meat came from, but it tasted good. We were pretty hungry at the time.'

'And the cook?' She still felt like being provocative. 'Did he have long dark eyelashes and scarlet lips and a dress torn in all the sexiest places?'

'The cook got zapped two days after. I buried him.'

'Oh!' The brutality of it shocked her, then she saw the pain in her husband's face. 'I'm sorry, Tom.'

'Well, you don't have to stop liking chili just because the guy who made it bought the ranch.'

'Let's talk about something else.' Her frivolity was gone. She would have liked to talk about her pregnancy now, but it seemed incongruous. She wanted the moment to be right. 'What kept you so late tonight?' she asked in an understanding tone. 'Who was raising hell?'

'The big man. You know the Soviets moved into a place in the Arctic called Svalbard? I happen to have been there, a few years back. Admiral King wants a position paper in a hurry.' Realizing she was already in a mood, he approached the subject gently, not mentioning that the Admiral required a lot more than just a briefing.

As he explained further she had a terrible premonition that he was going to be drawn into this, that just as she was preparing for the baby, fate would whirl him away to the other side of the world for no better reason than that he had been to this Arctic spot before and probably very few other Americans had. She was unable to disguise her concern.

'You won't be sent there, will you, Tom?'

'When it's full of Russians?' He laughed and shook his head. 'No, sweetheart, I don't think so. But I do have to make a trip to Oslo.'

'Oslo?' Her emotions immediately tautened. He always described his days in the Norwegian capital as the best time of his life.

'NATO's Northern Command is headquartered there. The guy in charge of the exercise is a four-star Britisher. I have to meet with his staff.'

She stiffened in her chair, straightening up and looking at him hard, all bantering finished. 'When do you go?'

He reached out and held her hand. 'Tomorrow, sweetheart. I'm sorry.'

'Tomorrow! Tom, you don't mean that?' She felt giddy and involuntarily tightened her grip on his hand. 'Do you have to go?'

'When the Admiral says move, sweetheart, we move. It won't be for long.' He could see she was really upset now, and tried to soothe her. 'I've been lucky to be away so little. This Special Operations kick can take guys almost any place.'

'And what am I supposed to feel? Tell me that, Tom Peterson. Where do I fit in? What do I get excited about? Being pregnant?'

'Being what?' Suddenly he understood, leaped to his feet and leaned across the table to embrace her. 'Are you serious, sweetheart? Why, that's wonderful, that's great! Why didn't you tell me before?'

'You never asked.' Now it was out she didn't know whether to laugh or cry. 'Tom, you do want a baby, don't you?'

'I certainly do.' He came round to hug her, pulling her up with his arms around her waist and swinging her off her feet. 'Nancy, that's great news, couldn't be better.'

'Be careful, you'll knock something over.'

'To hell with things, it's you I have to be careful of.' He kissed her again and at last put her gently down, still hugging her.

'I'm so glad you're pleased.' She didn't know what to say and found tears in her eyes. 'Tom, will you promise me one thing?'

'Anything you say.'

'Just be here when the baby's born. Don't let them send you anywhere. I'll need you.' She wanted to say, leave the Marine Corps, you've done enough. But she held it back.

He hesitated fractionally. 'That's one fact the Admiral will just have to face. I'll be here.' He kissed her again, trying to mask the impossibility of guaranteeing any such thing.

In the good old days, when army commanders were national heroes, the military enjoyed suitably resplendent headquarters. Even the red brick Washington home of the United States Marine Corps is now a national monument. But today the four-star general is likely to find himself out on some piece of real estate which is available primarily because no one else wants it. When the North Atlantic Treaty Organization was given notice to quit France by de Gaulle, back in 1966, the Belgians immediately offered to help. Down in southern Belgium, where the industrial city of Mons was badly in need of more jobs, there rose the prefabricated buildings for the Supreme Commander Allied Forces Europe. At the same time the diplomats and advisers who served the NATO Council left the impressive Palais de Chaillot overlooking the Seine for a quickly developed site at Evère, near Brussels airport.

Sandra never found the concrete structure of the Evère

complex anything but bleak, even though at the time she joined NATO young trees were growing to mask the severity of the entrance façade and a German ambassador had donated a large number of rose bushes to be planted between the long three-storey office blocks.

However, time and friendships softened the image. She grew accustomed to the security checks, the long passages sardonically known as *pas perdus* – wasted walking – and the incessant petty intrigue. There were over 2000 international and military staff at NATO, plus a further 1500 personnel on the diplomatic delegations of the fifteen member nations. It would have been extraordinary if the organization did not have something in common with the mythical Tower of Babel, even though French and English were the only official languages.

This morning, as she drove out along the broad airport road in the rush-hour traffic, Sandra's thoughts were not on her job. Her mind was still full of the night's memories. Erich had woken her early to make love again with a hard but short-lived intensity which left her unsatisfied yet happy at arousing such passion. Then, as was equally his habit, he let her make breakfast while he shaved, and she had dropped him off near his office. On an impulse, just before he got out of the car, he had asked, 'Are you free tomorrow? If there's good weather we could go flying.' Tomorrow was Saturday. She knew without consulting her diary that the weekend promised two staringly blank days, broken only by Sunday brunch with some office colleagues, pencilled in as a last resort. She had accepted Erich's suggestion avidly.

All right, she reflected as she stopped to show her pass to the guard at the entrance gates, so he has got me on a string, what does it matter? If I play hard to get I'll only lose him.

The policeman saluted and waved her through, past the flags of the NATO nations, hanging limp in the damp air, past the bronze monument of the four-pointed NATO star, to the car park in front of the building. She thought she scented tension the moment she was inside. Everyone was hurrying. Instinctively she did the same, walking fast along the yellow-carpeted *pas perdus* linking the centre block to the wing occupied by the International Military Staff. At the end of the wide corridor there was another security desk, manned by blue-uniformed civilians. A notice in English and French reminded employees going on leave to hand in their passes. It always made her think of Ursula, who had done no such thing before disappearing to East Germany in 1979. She had liked Ursula, even though she had been a spy. She showed her pass, pushed open the glass doors and hastened upstairs.

'Welcome to panic stations, miss,' said the British sergeant laconically as she entered the large room where she and two others worked. 'Emergency meeting of the Council at ten. In the Presentation Room. The General's doing a briefing.'

'On Svalbard?'

'Where else?' Sergeant Webb, whose duties were mainly clerical, was a small, dapper man with neatly brushed, thinning hair, a ready smile and a well-developed line of humorous banter. He would have made a good travelling salesman. As it was, he presided efficiently over the General's outer office, where he, Sandra and another secretary, Julia, handled the massive flow of paperwork. Julia was a blonde, self-confident to the point of brassiness, who wore a gold chain round her left ankle. She called Webb 'my sergeant' and he called her 'my secretary'. 'Now, miss,' he went on. 'Since my secretary isn't here yet, I'll have to ask you to put this little lot through the word processor. Hot from the Chief of Staff's hand, it is.'

'You might let me get my coat off!' Sandra protested. Somehow she never managed Julia's kind of repartee. As soon as she was ready she glanced at the draft. 'How many copies?'

'Fifty to be safe. It's a closed session so there's just the ambassadors plus one, but we're bound to need more.'

'Numbered, I suppose?' She noticed the heading NATO SECRET at the head of the paper.

'Numbered and signed for.'

'Off we go then.'

The word processor could type at a phenomenal 700 words a minute, once a text was stored in its computer. But it had one disadvantage. It gave off radiation which could be picked up through central-heating pipes, telephone lines, windowpanes and various other adjuncts of a normal office. So it had to be housed in a soundproofed, windowless room of its own, secure from electronic espionage. This was where Sandra took the text and began tapping it out, checking each line on the display screen above the keyboard. The brief began prosaically.

GEOGRAPHICAL SITUATION

The Svalbard archipelago is located between latitude 74° and 81°N and longitude 10° and 35°E. Approximately the size of Switzerland, its total 62,400 sq km lie in the Barents Sea at the southernmost edge of the permanent ice covering the Arctic Ocean. The flow of the Gulf Stream keeps the waters south of the archipelago ice-free for roughly half the year and makes the climate tolerable for human habitation. The largest of the islands is Spitsbergen, with the mining settlements of Longyearbyen, Pyramiden, Svea and Barentsburg. The remaining islands include . . .

At this point she lost interest in the meaning and concentrated simply on the correctness of her typing. However, the ending of the eleven pages recaptured her attention.

The strategic advantages accruing to the Soviet Union from the possession of Svalbard could thus be considerable. If allied to political aims, these could present the West with a most serious challenge.

So Erich was right, she thought. This might be the biggest story to come out of Brussels for years. For a moment she was actively curious about what would happen in the Council meeting, not that it would be a problem discovering.

If she really wanted to know she could probably find out from one of the interpreters. She considered doing this, then dismissed the idea. All the ambassadors ever did was talk. Today would be no different. She adjusted the processor to spew out the fifty copies and a few minutes later took them back for Sergeant Webb to sort into sets.

Normally the North Atlantic Council, the highest authority in the Alliance, met in the large Conference Room number one, with the motto ANIMUS IN CONSULENDO LIBER in high letters on the wall. But when the Council was taking military advice the fifteen member nations' ambassadors found it more convenient to be briefed in the Military Committee's own blue-curtained Presentation Room, which was effectively an operations centre with radio and telephone inputs, map displays and all the facilities a general needs to illustrate the points he is making. Indeed General Anderson's staff often called it the Situation Room and on this occasion they had assembled slides, satellite photographs and other illustrated material to supplement the written brief printed out by the word processor. Like most Americans General Anderson was a believer in visual aids, and as he stood to address the Council, seven rows of medal ribbons blazing above the breast pocket of his uniform tunic, he made continual use of them.

'Gentlemen, I am sure Ambassador Jacobsen of Norway will forgive me if I underline facts of which he is already aware.' The General raised his forefinger and a satellite photograph of northern Europe flashed up on the screen, the Arctic dome of the earth white, shading off to grey on the landmass of Scandinavia and Russia. A heavy black arrow superimposed itself on the photograph, pointing at the edge of the white. 'Svalbard is strategically important for five reasons. First, the archipelago lies astride the access routes for the Soviet Northern Fleet from the Kola Peninsula to the Atlantic.' The arrow moved down to rest on a thumb-shaped extension of Scandinavia, jutting east above Finland. 'The Kola Peninsula, gentlemen, contains the largest concentration of military power on the globe today. It is home base for the Soviet Northern Fleet, including 40 per cent of Russia's major fighting ships and 75 per cent of her most advanced ballistic missile submarines. Kola's airfields maintain 600 aircraft, including the fighter bombers of the 13th Tactical Air Force, long-range bombers and naval strike aircraft. The barracks house two motorized rifle divisions and a regiment of Marines.'

Anderson paused as a viewgraph flashed up on another screen listing Soviet military, naval and air resources in the area.

'Murmansk,' he went on, 'is also the Soviet Union's lifeline port for commercial shipping, the main ice-free harbour from which she has unobstructed access to the Atlantic.'

The ambassadors made a show of glancing at the lists. They all knew where this build-up of strategic information was leading.

'So, gentlemen. Svalbard is positioned to control both the Soviet Northern Fleet's surface access to the Atlantic, and vital commercial shipping routes. Thirdly, it lies in the heart of the nuclear missile submarines' operating areas. Therefore, the archipelago is crucial to any reinforcement of Europe from the United States and Canada because such reinforcement depends on control of the North Atlantic.' He paused again. 'The fourth factor is no less significant and relates to Longyearbyen airfield being in Soviet hands – '

'I do not agree.' Jacobsen's protest was immediate and taut.

'Our intelligence suggests, Ambassador, that it is.'

'There are still Norwegian officials administering the airfield.'

'Perhaps we can leave that point.' The Secretary-General, seated at the centre of the long horseshoe of tables, intervened diplomatically. 'General Anderson, please continue.'

'Thank you, sir. The fourth factor is that the shortest path any missile or bomber can take between the heartlands of the Soviet Union and the United States is the Great Circle route over Svalbard. If the Soviets were to install early-warning radar at Longyearbyen, it would place them 800 miles closer to the United States, greatly improving their defensive capabilities. If they militarized the field totally they could operate interceptor patrols against either B-52 or B-1 bombers, catching them before they could launch cruise missiles against Kola. They could also operate against our fleets and our merchant shipping.'

Again Anderson paused, while maps showing the radius of action of missiles and aircraft were displayed.

'Gentlemen,' he concluded, 'a Soviet takeover of Svalbard would achieve a major shift in the direct balance of power between the Soviet Union and the United States. In my estimation, this is the greatest threat we have faced since the 1962 Cuban missile crisis. I hope, gentlemen, that you can agree to the strongest possible response.'

'Thank you, General Anderson.' As chairman of the Council, the Secretary-General needed exceptional diplomatic skills. The ambassadors of all fifteen member nations spoke with an equal voice, even though there stood behind them widely varying military and political potential, ranging down to the Grand Duchy of Luxembourg, which possessed only a tiny army. Furthermore, the internal affairs of member countries were excluded from discussion. The ambassadors had only assembled today because Norway had requested them to. 'As I see it,' the Secretary-General continued gravely, 'this may be as much a test of our Alliance's political resolve as a military challenge.'

'Mr Secretary-General.' Jacobsen, a surprisingly young and energetic career diplomat, was determined to dominate the discussion he had asked for. 'Our view is that this is a response to increased American naval activity in the Barents Sea and the positioning of cruise missiles on United States aircraft carriers in the Atlantic. None the less, as an attempt to share Norwegian sovereignty, it is basically political.'

There was a constrained silence. Everyone present knew

how acutely the Alliance was being tried by the neutralist campaigns in Belgium, Denmark, Holland and Britain. The intense campaigning against cruise missiles in Europe had finally prompted the United States President to put a number aboard US carriers, from which aircraft could launch them over the ocean. This had provoked a barrage of protest from the Soviet Union and from left-wing activists. What no one had predicted was Soviet retaliation against NATO territory.

'I do not need to remind the Council of the background to our decision,' cut in the American Ambassador. 'We believe it was the only way to counterbalance the Soviet strategic missile programme. I also believe the Soviets now hope to split the Alliance because strategically Svalbard is a far greater threat to the United States than to Europe.' He turned to the Norwegian. 'Ambassador Jacobsen, may I ask what action your government proposes taking?'

This was the crucial question. In the background the interpreters poised themselves.

'In this same month, February, of 1978, our then Prime Minister, Mr Nordli, declared, and I quote him, "It is Norway's responsibility to govern Svalbard." This was after problems concerning the Soviet helicopter base there. Mr Nordli said that the Soviets had adopted a position on the Svalbard Treaty which we could not accept. The same is true today.'

General Anderson shifted angrily in his chair. What the hell were the Norwegians specifically going to do? Declare a national day of prayer?

'Apart from making our United Nations action,' Jacobsen continued, 'we are sending a ministerial delegation to Svalbard to assert Norway's rights.' He paused. 'The obvious question is whether we should ask for reinforcement of the forthcoming NATO exercise in northern Norway.' He paused again, though not for long enough to allow anyone else to suggest the answer. 'My government's view is that we should plan for a show of Allied support during

Polar Express, but the implementation of the plan must depend on Soviet reactions to our protests.' He glanced at Anderson. 'Our appreciation is that to date the Soviet move has no military content.'

Anderson was silent. He knew from the infrared satellite photographs now coming in that Longyearbyen airfield was already sealed off. The spy satellite pictures showed the barricades clearly. They also revealed bulky packing cases being unloaded from freight aircraft. But the Norwegians had men on the ground. This was not the time to dispute their reports.

After further discussion, the meeting broke up with agreement that both the Supreme Allied Command Europe and the Supreme Allied Command Atlantic should be ready to step up next month's exercise at short notice. Anything more, the Council felt, would be considered provocative by the Soviet Union. 'Norway,' Jacobsen had emphasized, 'can never be a party to a policy with any smell of aggressive intentions. We cannot risk undermining the Nordic balance.'

That afternoon consultation began between General Anderson's staff and officers down at Mons and in Norfolk, Virginia, where Admiral King was recalling absent staff from leave. Shortly before they were due to leave, both Sandra and Julia were asked by the Chief of Staff if they could work late.

'But Brigadier Curtis!' Julia wailed, her blue eyes wide. 'My boyfriend will kill me if I stand him up this evening. He's spent a fortune on theatre tickets. The Bolshoi are only giving three performances.'

'Very appropriate, in the circumstances,' said the Brigadier wryly. 'This is top priority, though.'

'I'll stay,' offered Sandra. 'I'm not doing anything in particular.'

'Darling, that's sweet of you. Enrico would go mad.' Julia smoothed back her long hair, as though already preparing for the night out.

'It's all right. You can stand in for me some time when you haven't a date.'

'That is very, very seldom.' Julia smiled glacially, stood up and reached for her coat. 'See you tomorrow, all.' She swept out.

The Brigadier laughed and shook his head. He had wispy red hair and a schoolboyish expression. 'Lucky Enrico. Thanks, Sandra. We'll try and get this lot through by seven.' He grinned again. 'The General's firing off orders like a machine gun.'

'Is there a crisis?'

'We think so.' The smile left his face. 'But if you'd heard the Norwegian Ambassador this morning you might wonder. His government is all set to tell the Soviets they're naughty boys and must go home. It was a wise man who said politics are too important to be left to politicians. Come on then, the sooner we get these signals off, the sooner we can go.'

In the end Sandra was not finished until eight and when she left she noticed that the building, normally all but deserted at this time, still had lights in many of the delegations' windows. She thought of Erich and wondered if he had written a story for his magazine. The answer came unexpectedly soon, only minutes after she reached her apartment. The phone rang and the voice on the line was his, slightly harsh and guttural.

'How did the day go then? You are back late.'

'Yes.' She was surprised. 'Are you at the Hilton?'

'They did not want me tonight.' He reacted quickly to her tone, seeking to reassure her. 'I rang to confirm that we meet tomorrow. I looked forward to that.'

'So do I, darling. Very much.' It crossed her mind to ask him round, if he wasn't doing anything, but she hesitated. She didn't want to throw herself at him. She ought to learn some of Julia's tricks. 'Just hang on a moment while I get comfortable, will you?' She put the phone down, settled herself on the sofa, then resumed the conversation. 'And how was your day?'

'Frustrating. Absolutely frustrating.' He took the cue instantly. 'You know what communiqué the NATO issued?'

'I didn't see it.'

'I will read it. Listen. "The NATO Council met in closed session and considered the serious issues which arise from the Soviet interpretation of the Svalbard Treaty. The Allies will remain in close consultation." That is all. What can I make of that, for God's sake? The Press Office refuses to give guidance. Is this nonsense what they keep you working late for?'

'Darling, I don't have anything to do with the Council.'

'But they cannot be doing nothing!' He was half shouting down the telephone. 'What can I say to the magazine? That I failed to find out?'

Hearing the despair in his voice, she tried to help. 'The only thing I heard was that the Norwegians will tell the Russians to leave.'

'Without military threat?'

Suddenly Sandra felt alarmed. Even as a secretary she understood how dangerous it would be if newspapers suggested NATO was taking military action. She ought to dispel that idea at once. 'Definitely,' she said. 'That much I do know. The only military thing going on is next month's exercise.'

'Which will now be expanded?'

'I think so.' She felt caught by her ignorance. 'Really, darling, it takes them weeks to decide anything.' She sat up straight, as though the physical move would somehow be felt down the telephone. 'Listen, do you want me to pick you up tomorrow morning and shall I book the Cessna?'

'Yes. Yes, of course.' Reluctantly he accepted the change of subject. 'The morning is best, then we can have wine with our lunch.'

The mention of wine made her tremble. In the language of their affair wine with lunch signalled afternoon love-making, uninhibited and delicious. She felt so excited she could hardly speak. 'I'll pick you up at ten,' she managed to say, then put the phone down with a mixture of relief and sadness. She wanted desperately to ask him round, but he was obviously in a state over his article and she was afraid of a quarrel.

In the hotel room where he was sitting, Erich looked sharply at the man with him, who had been listening through an extension earphone clamped to the side of the instrument. 'That is probably the truth,' he said in German, stroking his chin. 'I shall find out more tomorrow.'

'And we must be in more frequent communication,' the man replied, carefully unfastening the extension wire and putting it in his pocket. 'Much more frequent communication.' He shifted his position, trying to sit upright on the side of the sagging bed. 'Your efforts are beginning to pay off.'

'I hope so,' Erich agreed frankly. 'I damn well hope so. If they'd told me three years ago that I was leaving the Air Force to become a gigolo, I'd have refused.'

'It is a privilege to work for the Stasi,' the man said stonily. 'You are lucky. If you succeed with this project promotion will be certain. Most men would do it for nothing. Don't you enjoy a good lay?' He sounded envious himself.

'Naturally.' It was Erich's turn to bristle. 'But this is like making love to a camel.'

'Don't exaggerate.' The man rose and picked up his briefcase. 'Just concentrate on getting the information. Our Soviet friends want daily reports, hourly if necessary. We haven't been in a position to give them such service for a long time. The *Aktuelle Nachrichten* editor will provide a regular stream of journalistic queries for your cover. This is the chance of a lifetime, Braun. Make the most of it.'

None the less, as he walked back to his rented room through the wet and deserted streets, Erich wished he had not given up flying jets for the allegedly more exciting life of an agent.

The East German Security Service, the *Staatssicherheits-*

dienst, who were Braun's true employers, had devoted much long-term effort to establishing a background for him. The identity card issued to a former Munich resident by the *Polizei Praesidium* in that famous city and which Erich now carried was completely genuine. Its original owner had been imprudent enough to get into trouble while visiting relatives in the East and Erich, in a manner of speaking, had inherited his name. In turn the identity card enabled him to obtain a valid West German passport. The left-wing magazine which employed him as a stringer in Brussels was equally genuine, though one particular member of its staff was more committed to the socialist cause than the others. In his articles Erich deliberately took a moderate political line. If he could make the transition to *Der Spiegel* his careful metamorphosis would be complete.

His case officer in East Berlin, monitoring Erich's development from a paper-laden desk in the Stasi's ugly concrete headquarters in the Normannenstrasse, was pleased. Erich had come a long way since the political officer of his East German Air Force squadron had quietly suggested there might be a future for him in another area of government activity. Indeed the Stasi was finding his success almost embarrassing. The KGB was now breathing down their neck for more reports.

But for Erich himself there were other problems. He was worried in case Sandra exasperated him to the point where he lost his temper. He was already starting to hate her.

3

As the plane taxied out, Tom Peterson wondered how far the Russian general commanding the Warsaw Pact forces ever trusted his East European allies. Not that it made a damn of difference in Svalbard. Only the Russians themselves were involved, whereas on the NATO side there were Norway, Britain, Denmark, Holland, Iceland, the US itself . . . shit! The conclusion Peterson had reached after three days in the northern European headquarters at Kolsas outside Oslo, where the operations room was dug deep into a mountainside, was that potentially this was the biggest can of worms he had ever seen. The prospect of NATO agreeing on action seemed remote. He wasn't convinced he was going to learn much from visiting the Polar Express exercise either, though Admiral King had insisted on his doing so.

Their destination, the airfield of Bardufoss, was far up in the Arctic Circle. On the way his companion, a burly British lieutenant colonel called McDermid, who wore the distinctive parachute wings of the Special Air Service high on his sleeve, argued about the situation as he saw it.

'So how do you think we should winkle the Soviets out?' he asked.

'Fire a nuclear warning shot, maybe.'

'A nuke?' McDermid nodded. 'The Arctic's as good a place as any to test that theory.'

'Wouldn't fry anything more than a few bears,' Peterson went on encouraged.

'The only snag – ' McDermid's eyes were impish in his thickly fleshed face ' – is that the Norwegians would never agree. They are totally opposed to the first use of nuclear weapons.' He reached for a map and spread it across the narrow table. 'And suppose the Soviets took it as a challenge? They could roll up northern Norway in days.' He stabbed the map with his forefinger. 'Airborne assaults on the airfields, marine landings. Their armour would pour through Finland. NATO's northern flank would have been turned.' He jerked a thumb at the window. 'Would NATO go to war for this snow-bound wilderness? Would your President risk a nuclear attack on New York for it?'

'The Treaty says he would.'

'I don't believe so.' McDermid leant back in his seat and intertwined his fingers, almost as if praying. 'Try being the devil's advocate. Suppose the Soviet Union really feels as hemmed in and threatened as its leaders claim? What if all they actually want is to secure the North Cape sea route? Do we go nuclear to stop that?'

'I guess not.'

'That's certainly our thinking here. We have to find a way to persuade the Soviets to leave. If we don't, and they keep Svalbard, they'll have proved the Alliance can't defend Norwegian sovereignty. The next step would be a Soviet-sponsored freedom from aggression pact for Scandinavia. If Norway went neutral, they'd have turned our flank without bloodshed.' He laughed sardonically. 'My general shares Admiral King's view. We have to take steps to remove whatever the Russians instal on Svalbard, with or without Norwegian help. That's why he's keen for you to meet Simon Weston. He's a commando with some useful ideas. Don't be put off by his appearance. He's tougher than he looks.'

That evening there was a formal dinner with the Norwegians at an officers' mess near Bardufoss. But Peterson's hopes of further information went unfulfilled. The nearest anyone got to mentioning Svalbard was a formal speech welcoming exercise Polar Express as a demonstration of NATO solidarity at this difficult time.

Next morning he and the SAS colonel were flown to the exercise by a Norwegian Air Force helicopter. It was a Huey.

'Takes me back,' commented Peterson as they strapped themselves in. 'These birds were our friends.'

'In Vietnam, you mean?'

'Yeah. Heard one of these coming and we were sure glad.' The memory was like a sudden ache in a wound you thought was healed, or the picture of a woman long forgotten. Heat, sweat, the bullying optimism of the captain who always wanted to make one hill farther. Humping through that goddamn jungle in 110 degrees of heat and soaking humidity, never sure where the Cong were. How in hell could that memory stab you in the gut?

The Huey swayed into the air and raced forward, flashing over snow-laden trees and the flat coil of a river. A large deer stood motionless in the open.

'Elk!' shouted the SAS colonel. 'Bit different from Vietnam, eh?'

'Yeah.'

The noise precluded conversation. Anyway the sight of that solitary elk threw up images from the movie *The Deerhunter* and a host of other thoughts followed. He had been an ordinary Marine in Vietnam, eighteen years old, thrown into action by that damn South Carolina judge. He reckoned now he could have gone either of two ways after that. Done his two years and emerged resentful at a charge which had been essentially vindictive. Criminal assault. Shit! The guy had been pestering his mother, he deserved to have hell beaten out of him. What Peterson hadn't known was that the slob had friends in the local police department.

The combination of the trial and Vietnam might easily have turned Peterson the wrong way. In fact it hadn't. Out there in the jungle he had come to terms with what he could and could not demand of himself. So for him it had been like cauterizing a wound. But the wider issues of the war itself were something different, something on which his perspective had changed as he grew older. The people back in the world, the ordinary American citizens, had wanted to know what had gone wrong. Why hadn't the richest and strongest nation on earth won the war? He knew it had been unwinnable. But Vietnam was also a place where men helped each other with no questions asked, which was one hell of a lot more than one could say of America. So he had stayed with the Green Machine. But equally he never again wanted to be the fall guy for a wrong political decision. Not if he could help it. And that brought him back to the present. Was this going to turn into the biggest can of worms since Vietnam?

The Huey circled, rocked in a blast of wind. Below, dug into the snow, were tents under white tufted camouflage nets. Men in white coveralls were moving slowly on snowshoes, rifles slung over their shoulders. A line of M-60 tanks were half off the nearby road. The Marine Amphibious Unit looked pretty much like a sitting duck.

'How long d'you want with them?' asked the colonel, after the Huey had touched down.

'Couple of hours.'

'We'll be back.'

A Marine major, a black woollen cap on his head, came up and saluted. The Huey took off again, momentarily blinding them.

'How are things?' Peterson asked. He wanted this visit to be as informal as possible.

By the time the Huey returned Peterson knew his first impression had been broadly correct. The grunts cracked the usual jokes, the tanks had the usual slogans painted on them, like 'Tracks of Death' and 'Have gun, will travel'. But the training in Minnesota had just not prepared the unit for this extreme cold. Even the snow was different here, they said. Ninety per cent of their effort was going into survival.

He recalled McDermid's words: 'Russian conscripts spend two, maybe three winters in the north. They're used to tough weather. We have a problem matching their acclimatization.' The Marine Amphibious Unit had enthusiasm, but it was far from ready for an Arctic war.

Later the SAS colonel said, 'The worst enemy is the climate. Except for those who have been here many times, our own men aren't up to fighting efficiently either. On paper this all looks great: 12000 troops, aircraft, ships. But essentially it's flag waving. The real resistance would come from the Norwegians themselves. As for Svalbard – '

'I don't see the Marine Corps landing there.'

'Let's go and find Simon Weston. He's somewhere up the Stor fjord. This time we'll drive. Makes our arrival less noticeable.'

The snow-covered road wound along the contours of the fjord, past isolated houses and tiny villages. After some thirty kilometres, the driver stopped where a wooden jetty projected into the water. Alongside the jetty was a small trawler. Peterson noted her name, *Northern Light*. The Colonel led him down, their boots crunching the snow, and shouted, 'Hallo there!'

A thickset man in a once white sweater emerged from the wheelhouse, took the pipe from his mouth and greeted them.

'How are you, Ted? Welcome aboard.' He shook hands. 'Simon Weston's the name. You're just in time for lunch.' He walked forward and opened a hatch in the deck. 'Sergeant Millar. Our guests have arrived.' He turned back to Peterson. 'I assume you'd like the guided tour first?'

'Any time.' Peterson felt obliged to disguise his surprise. Had the British brought him 700 miles to see this? He tried to make appropriate remarks.

The *Northern Light* was a traditional fishing boat, wooden-hulled, the planks varnished except where a white band was painted all the way round. Her number, F 192 H, was conspicuous on both sides of the bow. She was about sixty feet long. A white wheelhouse with six windows in it stood stark and unstreamlined behind a tall mast with a railing round its roof and a clutter of equipment. Peterson noticed a searchlight and the bar-shaped arm of a radar scanner. Behind the wheelhouse was a cabin. There was ice on the rigging. All in all, the *Northern Light* was any landsman's idea of a trawler, the kind that ventured out to bring back cod for the fish fingers sold in a thousand supermarkets.

'It's all pretty standard,' said Weston, pride edging into his voice. 'Except for the gear we've added. The real joy is the hold – 600 cubic feet.' He led the way forward, opened a hatch and descended. It was fitted with wooden bunks. 'We can take sixteen men and their kit. The rear hold's fireproofed for ammunition storage. It's all ventilated, heated, the lot.'

'How fast does she go?' Peterson asked as they emerged again.

'Slow. Ten knots. But there are two thousand like her in Scandinavian waters. She's a normal sight anywhere from the Barents Sea to the Baltic. Shall we eat?'

The sergeant had cooked sausages, fried potatoes and eggs in the tiny galley off the cabin. They sat on leather-cloth benches which became bunks at night.

'Frankly,' said Peterson, 'I can think of other ways of getting a raiding party ashore.'

'Absolutely.' Weston's reaction was immediate and open. 'Pop them up out of a submarine, free-fall from an aircraft. But the amount of kit they can take is limited. Insertion by chopper's better, except that like the sub it is liable to be detected. May I tell you something, Colonel? The predecessors of these boats sailed to and from Norway throughout the Nazi occupation. Ideas aren't necessarily bad because they're old. At all events, I'm trying this one out against the Norwegian Navy this week.'

'Do they know?'

'Like hell they do! What's the point of being clandestine

if you tell everyone?' Weston grinned broadly, like a schoolboy with a stunning new scheme. 'If we can pass ourselves off as fishermen to the Norwegians, I reckon we'd have no trouble persuading Russians. Which is the name of the game, after all.'

'If we can't, this boat'll be as much use as a pogo stick in a swamp,' cut in the sergeant. He had an earthily vivid turn of phrase.

By the time Peterson left he was partially convinced, though the application of the *Northern Light* to Svalbard seemed remote. The ice wouldn't melt up there until June. Everything would be over by then.

In Oslo he found a letter from Nancy waiting for him. After some paragraphs of domestic news there came a punch line. 'If you're not back soon, I'm coming across myself. What's so fantastic about Norway anyhow?'

He cabled at once. 'Norway deep-frozen. On my way. Love you, Tom.' He wished he could have worked out something snappier.

Nancy Peterson knew Tom was a lousy correspondent at the best of times and she had written querying his return before she could possibly have expected a letter from him anyway. She told herself she was merely being perverse and that this worrying must be a symptom of pregnancy which reason could overcome. She tried to distract herself by reading an aged copy of Dr Spock her mother had sent. It was no good. She remained tetchy and restless. She just knew Tom was going to be up to his neck in this Svalbard thing and she watched the television news more often than usual.

The programmes did nothing to cheer her. The first flurry of over-excited comment by defence analysts had been quickly followed by a blunt denunciation of the Soviet action from the Secretary of State.

'The message coming from the White House,' one well-known broadcaster explained, 'is that Soviet occupation of Svalbard is unacceptable to the United States. A similar attitude is being taken by almost all signatory nations of the Svalbard Treaty and maximum diplomatic pressure is being put on the Russians to withdraw. If it fails economic sanctions appear likely, though they have never worked too well in the past. Grain sales embargoes hit US farmers as badly as they hit Russia.'

Then, after Tom had been gone four days, came the news that the Norwegian government was sending a ministerial delegation to Longyearbyen to reassert Norwegian sovereignty. 'The hope here in Washington,' remarked the commentator, 'is that this legitimate challenge will force the Soviet Union to back down.' Evidently the Norwegians were encouraged by the strength of American and British support for their claim.

On the other hand Nancy found people she met, her fellow teachers especially, felt that Svalbard was one hell of a way off and basically a problem for the Norwegians.

'We went out on a limb with the British over that goddamn Falklands invasion,' one man said. 'We should be more careful.'

'Sure, but this is different,' argued another. 'This is a threat to us.'

Nancy listened and for once didn't feel like stating a position. Intellectually, she knew America had to react. Emotionally, she was terrified. When Tom's cable was delivered she found herself crying from relief. The longer Norway stayed deep-frozen, the better.

For Frederik Folvik the days since the Soviet takeover had been bewildering. When he telephoned his mother in Oslo he told her he could not believe such things were happening. 'They have cordoned off the airfield,' he said. 'But we still work the same shifts. The only difference is there's always a Russian with me.'

At this point a number of clicks on the otherwise distinct telephone connection reminded him that conversations were being monitored. Makarov had openly told him so, as if giving a friendly warning. Folvik shifted off the subject of the airfield.

'The mining goes on normally, the House is open – you know, Mother, the social centre. There's no interference with our daily life, though Russians have moved into the Sysselmann's office and share it with him.' He told her a few more details, yet failed to convey the curious atmosphere: why should some of the local communists object so vocally, for example? He was relieved when the conversation ended. He didn't like the line being tapped.

Eventually he did manage to pin down his thoughts more accurately. What gave the situation its peculiar feel was that although the Russians were demonstrably able to give any order they wished, now having over two hundred armed police to enforce it, they were behaving as though day-to-day decisions were by joint agreement. It was an elaborate charade, which only collapsed where certain matters were concerned, like the airfield and communications with Norway. Inevitably the man who personified the play-acting for Folvik was Makarov, the Aeroflot manager.

Take today, for instance. It was Tuesday and Folvik had driven out to the airfield as usual in his old Volvo in time for the afternoon shift. The Russian guards stationed at the newly built checkpoint near the airfield perimeter had examined his identity card and saluted as they raised the barrier. They had never saluted him before. Furthermore, there was a lone Norwegian policeman with them, who had smiled sourly, as though aware that he was only there for show. But the real surprise came after he had left the car on some waste ground near the long, low wooden building of the government hotel – parking was no longer permitted by the tower itself and the two dozen rooms in the hotel had been taken over by the Soviet border police officers. Folvik had heard shouted military commands as he clambered out into the snow. When he got closer he saw a contingent of border police being drilled on the tarmac in front of the tower, which had been cleared. A soldier in a peaked cap, whom Folvik assumed to be a warrant officer, was shouting the orders and there, standing to one side but evidently the senior officer present, was Makarov in his heavy black overcoat and astrakhan hat.

My problem, Folvik told himself as he climbed the tower stairs after submitting to a further security check, is that I take people at their face value. The previous Aeroflot manager had been in Africa and had shown them all snapshots of elephants. He had seemed a straightforward airline official. But he had been replaced four months ago with minimal warning by Makarov, who was entirely different. Self-controlled, hard-minded. Folvik thought of the moment last Wednesday when the Tupolev had so nearly failed to stop in time and how Makarov had insisted the landing was without fault. 'We have most experienced pilots.' Folvik felt sure that if the plane had crashed he would have instantly blamed the airport authorities.

Normally, in the seemingly remote past of a week ago, Folvik would have chatted for some minutes to the controller from whom he was taking over. Now, with an aggravatingly arrogant young Russian present as well, a Russian who spoke excellent Norwegian, their conversation was constrained. But Folvik did ask what was happening outside.

'A delegation from Oslo is due in fifty minutes,' explained the colleague. 'An SAS DC-9. The details are on the telex.'

'From Oslo?' One of the big unanswered questions of the last few days was whether Tuesday's scheduled flight from Tromso would be allowed in. Folvik stared down from the wide windows at the parade below. 'What are they going to do?' he asked apprehensively.

'It is a guard of honour,' cut in the Russian. 'Naturally the Comrade Colonel will welcome them.'

'Makarov, you mean?'

The Russian flinched, as if caught out, and hastily corrected himself. 'Comrade Makarov is in charge of transport and communications. It is his duty to be present.' He turned away.

Half an hour later the pilot of the DC-9 made a radio call to report his approach. The young Russian glanced at Folvik, then left the room, his footsteps sounding heavy as he hurried down the stairs. Folvik saw him run out and report to Makarov, who immediately gave an order to the sergeant major. The police moved stiffly back, as if half frozen inside their greatcoats. Left to himself, Folvik decided that he must warn the Governor whatever the consequences, and rang through on the direct line. An unfamiliar voice told him the Sysselmann was already on his way. A couple of minutes later he heard cars draw up outside and looking down again from the side windows he saw both the Norwegian Governor and the so-called Soviet Minister for Svalbard arrive. They vanished, presumably into the terminal building.

Folvik duly informed the DC-9 crew that the runway was free for landing and the airliner came in, throwing up whirls of freshly fallen snow as it touched down.

The events which followed amazed him. Remembering last Wednesday, he fully expected the Norwegian government delegation to be arrested as they descended the aircraft steps. It would be in character with Makarov's past actions for the young Russian's explanations of a welcome to be cynical mockery. Yet no sooner had the DC-9 halted in front of the tower than four police ran out, unrolling a length of red carpet across the tarmac, already flurried with windblown snow despite the earlier clearing. These police stamped to attention as the first figures appeared in the plane's doorway. At the end of the carpet the Soviet Minister and Makarov stood almost equally rigid as the roly-poly figure of the Minister of the Environment came down, clutching her fur coat to her with one hand, the Justice Minister following. As her foot touched the carpet, a roared command from the sergeant major brought the assembled police to life, their arms and legs moving with the jerky precision of clockwork dolls as they presented arms.

Even at 50 metres' distance Folvik could see the surprise on the Minister's face. She smiled graciously, shook hands with the two Soviet officials, then she and the Minister of Justice solemnly inspected the guard of honour while their group of aides waited. Five minutes later they were being ushered to the waiting cars and disappeared from view. Still dumbfounded, Folvik watched the police briskly roll up the carpet again, leaving the crew no such luxury for their disembarkation. Then the voice of the young Russian jerked him back from the window.

'On Comrade Makarov's orders the airfield is now closed to all traffic except helicopters from Barentsburg. Is that understood?'

Folvik swung round, inspired to momentary defiance by the obsequious reception given the Norwegian officials. 'If Makarov wants that, he should consult us first.'

The young man faced him. 'You have seen that he is busy. You argue too much, my friend. I repeat, the airfield is now closed.'

'An excellent parade, Viktor Mikhailovitch. Precisely what we required.' The Soviet Minister, General Stolypin, congratulated Makarov as they were driven back from the airport to the town. In front of them the two Norwegian ministers were ensconced in an official Volvo with the Sysselmann. It was a 4-kilometre ride close to the shore, past coal storage buildings and the jetty which looked grimed even in the snow and the twilight. But Stolypin paid little attention to the surroundings. His mind was on the mechanics of the coming meeting. 'If these Norwegians decide to stay overnight,' he said, 'a suitable dinner party must be arranged. Remember, they are officially friends until they declare themselves enemies.'

'I will suggest it to the Governor's assistant.' Makarov

deliberated this briefly. Facilities in Longyearbyen were limited. In his guise as Aeroflot manager he had a house, but it was hardly suitable. 'It will be better if they invite us,' he said.

Stolypin grunted assent. That was a shrewd observation. Makarov had political sense as well as engineering expertise. On paper, the Colonel's record was outstanding. However, as a career officer who had moved into government and become a member of the Central Committee of the Communist Party of the Soviet Union, from which the ruling Politburo was drawn, Stolypin knew that paper qualifications were seldom enough. An assignment as tricky as this one demanded quick thinking and personality as well. Many months of planning had gone into the outwardly precipitate Svalbard move, but now that it had been made, its success hinged on himself and Makarov.

The two men, whose roles had been so crisply defined beforehand, were far from similar. The General, rather small in stature, but with a calculating and devious personality, would play the Politburo's game with the Sysselmann, maintaining the fiction of cooperation for as long as possible. Makarov, fifteen years younger, more physically forceful and straightforwardly determined, would organize the construction of the surveillance radar up on the snowblown Plata Berget mountain and the early warning system on the Isfjord headland.

Chance is always the taskmistress of the professional soldier. The career structure can take him a long way, provided he does the right things competently, but only chance can throw up the moment of opportunity which catapults a man to glory. Makarov had belonged when a schoolboy to the DOSAF – the voluntary society for the support of the army, navy and air force – and learned to shoot, swim, wrestle and drive before he went to the university and took his engineering degree. His father had been an army major in the Great Patriotic War of 1940–45 and because the Party encourages the development of military families, it had been natural for him to go on to the officers' school. For ten years he did the right things, most particularly becoming a Party member, which was a strong factor in his being transferred to the Border Troops, who had to be politically reliable. Then, as a major, he was sent to advise the army of the Marxist government in Angola and it was there, in the hot, dusty thorn scrub of southern Africa, that chance threw up his moment of opportunity.

Most of the fighting in a triangular battle against the South African army and the guerrillas of Jonas Savimbi was being done by Cubans, surrogates for the Red Army in Angola. The Politburo had no wish to become involved in an African Vietnam. None the less, a few hundred Russian military advisers were there and Makarov's job was to improve the communications along savagely bad dirt roads and across rivers which could be dry beds of sand and rock one hour and destructive torrents the next. As if this was not bad enough, the government's grip on the territory was so tenuous that the only truly safe method of transport to the south was by air.

One broiling, cloudless day a senior KGB general accompanied the Angolan Defence Minister to inspect Makarov's project. While they were looking at a half-finished bridge, a South African long-range patrol appeared without warning in a posse of helicopters to blow it up, while the scream of supporting jets sent most of the African soldiers racing for cover in the bush. Makarov personally rallied a company of Cubans, though they could barely understand a word of his shouts, and fought off the attack with such determination that the South Africans withdrew, leaving one wrecked helicopter behind as they retreated in clouds of smoke and dust. Although the Angolan Minister was wounded, a grave humiliation was averted. The general, deeply thankful at not being killed or, worse, taken prisoner, arranged both accelerated promotion and the order of Aleksandr Nevskii for Makarov. It was the turning point in his career and led directly to this far more crucial assignment at the other end of the world in Svalbard.

The car reached the straggling settlement of Longyearbyen, passing the harbour and going up a winding road to the Sysselmann's office, a large wooden building overlooking the waters of the fjord, now just visible as a pallid area of snow and ice. Snow crusted the sloping roof of the office and nearby a line of coal buckets hung from an overhead conveyor line, a reminder that the town's existence was completely geared to mining. A small crowd of locals had gathered by the driveway, clad in lumberjackets and parkas, fur hats with ear flaps on their heads. Two blue-uniformed Norwegian policemen held them back, while the other two stood on the veranda entrance of the building, together with two Russian police. The crowd was silent, suspicious, peering at the car.

'We may have unexpected trouble here,' commented Stolypin shortly. 'They are less friendly than I anticipated.'

Once inside, the two Russians were joined by other officials and escorted through to the conference room, where the windows were curtained. A long table had chairs ranged down each side, so that the delegations could face each other, and miniature flags, as well as note pads and pencils. They were shown to their places by the Sysselmann, previously a senior civil servant in the Foreign Ministry. He had one of those angular faces which characterized the old Vikings and are still strongly evident in Scandinavia today, with a neat beard and thick hair cut almost as short as the Russian's. He had dressed with formality for the occasion and was insisting on his rights in minor ways, like inviting the Russians to be seated. He was well aware how acute a cognizance all Soviet officials take of such details. But the talking would be done by the Minister of Justice from Oslo, who was administratively responsible for Svalbard.

'We have come,' the Justice Minister began uncompromisingly, 'to insist on the recognition of Norwegian sovereignty in the territory of Svalbard and the withdrawal of all Soviet personnel from Longyearbyen, excepting the authorized five members of the Aeroflot ground-control team.'

Stolypin nodded and spoke slowly to his interpreter, who then read out a prepared statement, prefacing it with an explanation that the original of the document had already been handed to the Norwegian Embassy in Moscow. With many circumlocutions, it invoked the agreement made between the Soviet and Norwegian governments in 1945 during discussions over joint rule in Svalbard, reminding the Norwegian government that after the Soviet Union relinquished its claim to Bear Island, the Norwegians agreed to the defence of Svalbard being viewed as the joint responsibility of the two countries. 'Until the recent aggressive moves by the American imperialists in the North Atlantic and Barents Sea,' the statement continued, 'there has been no requirement to implement this agreement.' However, the positioning of cruise missiles aboard a United States ship in these waters posed an imminent threat to the security of Svalbard. Finally, the document went on to claim that economic development was the other arm of Svalbard's defence and that since Norway could not contribute enough resources to such development, the Soviet Union wished to add to them under Article Three of the International Treaty.

This part of the claim, Makarov knew, was intended purely to confuse the issue and to prolong the wrangling. If taken to its logical conclusion a dispute over the interpretation of the Treaty could end up in the International Court of Justice at the Hague, from which a judgment might not emerge for five years.

Wisely, the Norwegian Justice Minister concentrated on denouncing the 1945 defence agreement on the grounds that it had never been ratified by the Storting, the Norwegian parliament, and that the Treaty anyway forbade the military use of Svalbard.

'We cannot be held responsible for the failures of your parliament,' Stolypin retorted. 'In our eyes an agreement is an agreement. The King of Norway may have forgotten the Nazi attacks on Svalbard during the Great Patriotic War. We have not. Germany was a signatory of the Treaty. Did that prevent the bombing of this settlement and the establishment of Nazi radio stations on the outlying territory? Is the United States any more likely to be inhibited by being a signatory? The Union of Soviet Socialist Republics cannot accept that the archipelago of Svalbard is safe from attack until the warships of the United States fleet withdraw from these waters.'

Not elegantly phrased, thought Makarov, but who cares? The challenge was now in the open. Either Norway persuades the United States to back down and withdraw the shipborne cruise missiles, or the Soviet Union stays in Svalbard.

'I emphasize, of course,' Stolypin continued, 'that any defensive measures taken by the Soviet Union here will be subject to full consultation with the Norwegian representatives. Our only interest is in the safeguarding of our mutual interests as peace-loving nations and in the development of natural resources for the benefit of our two countries.' He then inserted a neat red herring into the negotiations. 'We shall of course respect all statutory Norwegian regulations regarding the environment and the national parks.'

This gave the opening that the Environment Minister had wanted. Her half-hour of questions, all answered with the maximum courtesy and emphatic assurances, slightly defused the tension. But there was in any case little the Justice Minister could do. With four policemen at the Governor's disposal, plus ten they had brought in the plane, set against 200 available to Stolypin, there could be no threat of physical force. Nor, when the Russian announced his intention of sending frontier police to the other Norwegian settlements of Ny Alesund and Svea, plus action 'to safeguard' the radio station at the mouth of the Isfjord, could the Justice Minister do more than formally protest.

Although the Norwegian delegation accepted refreshments, they declined the dinner invitation and departed from the airfield only four hours after their arrival, leaving behind the ten additional police. Stolypin and Makarov stood ceremoniously on the tarmac as the guard of honour again presented arms. When the aircraft door had closed, Makarov looked up, beyond the airfield, to the great bulk of the Plata Berget, its flat summit visible as a hard white line against the dark sky.

'We are coming,' he said softly to himself.

Stolypin intercepted the glance but the whine of the jet's engines overwhelmed the words and he misinterpreted Makarov's expression. 'You are anticipating problems?' he asked.

'The sun does not rise for another week, Comrade General. I must survey the site in daylight. That is all.' Despite his four-month residence in the guise of Aeroflot manager, Makarov had only been able to get up to Plata Berget once, on a snow scooter, by moonlight.

'The radar experts from the Air Defence Command will be here shortly,' said Stolypin gruffly. 'And do not worry about NATO. Others are occupied with those contingencies. Our job is here.' He permitted himself a sly smile as the Norwegian jet lined up at the end of the runway. 'We shall play the game of cooperation with them up to the last moment. We shall argue and bargain until we are ready and they will remain confused. Democracy has always been its own worst enemy.'

That evening Makarov wrote to his wife Svetlana at home in Leningrad, where they enjoyed the huge luxury of occupying three rooms not shared by any other family. As a major Makarov had been employed in the Leningrad Military District and they had been allowed to retain the apartment during his sojourns abroad. This, too, was a mark of official approval, though in private Svetlana could be caustic about the price they paid for it in enforced separations. Despite her own civil service job, her great ambition was to be able to join her husband. But he seemed destined for a series of hardship posts – hellholes, she called them – where a wife was superfluous and unwelcome. She could hardly have accompanied him to Angola. But she cherished ambitions concerning Svalbard. Surely once the territory was Russian, she could come? Secret as his mission was, he had confided to her that the realities of Russians outnumbering Norwegians by two to one would soon be recognized.

So he wrote to her in the somewhat stilted phrases born of his education at the military academy, caution overlaying all he said. There was a third party to every letter home and, contrary to those jokes about the marriage bed being shared with Lenin 'because he is always with us', the third party in this case was the censor. The Border Police Troops might be a part of the KGB's empire, but the KGB itself still watched them. So Makarov was circumspect and only referred to General Stolypin in an ostensibly complimentary way, knowing his wife would read between the lines.

'My darling,' he began. 'By now you must have heard on the radio of the new situation here. Our motherland's rights are at last being recognized. More exactly,' – he searched for comparisons and remembered a phrase from a novel – 'the place has fallen into our hands like a ripe fig when the tree is shaken. These Norwegians are no match for the General, who has thanked me for preparing the ground here. In fact it is a relief to be back in uniform. Our next tasks will be difficult but straightforward.' That would give Svetlana a clear indication that he was back to military work.

He continued with inquiries for her health and the progress at school of their eleven-year-old son, Alexei. 'Does he still want to follow me into the Army? It's not a bad life, all in all.' That was an understatement. He himself might be a troubleshooter, often sent to distant places to sort out problems, but he and his family benefited. The 'certificate rubles' he earned abroad could be used as hard currency to buy imported goods in the special *Beryozka* shops. As an army officer his family could use the private facilities in the basement of the Army and Navy store on Moscow's 5 Kalinin Prospekt. They had a host of privileges not available to ordinary Soviet citizens. 'Yes,' he went on, almost reflectively, 'the initial training is hard but it's worth it in the end.'

Then it occurred to him that he was giving an unduly materialistic impression. Basically he enjoyed the life of a professional soldier.

'I can't give details, naturally, but the task here will be extremely challenging. One reward, I am sure, is that when the summer comes you will be able to visit me. Anyway this job will certainly be over by the autumn.' He concluded by reiterating how much he missed her and signed off with the affectionate diminutive of his name, Viki.

As was his practice he did not seal the letter immediately, but left time for second thoughts in the morning. However, nothing disturbed his sleep. Everything connected with the Svalbard takeover was going extraordinarily well.

The narrow valley where the American Longyear staked his coal concession in 1905, and where Makarov's police now patrolled, was scarcely two and a half miles long. It sloped up from the Isfjord between precipitous mountains into which the mine shafts drove horizontally and was locked in at the far end by a further peak with a glacier on each flank. A stream, frozen in winter, flowed down from the glaciers and divided the valley. In consequence a road ran up each side, one coming from the airport and quay, the other from a longer valley called Adventdalen. The Governor's office and the church were on the airport road, while most of the small family houses and miners' accommodation blocks were on the Adventdalen side.

Two days after the ministerial delegation's visit, a thirty-one-year-old mining surveyor called Paul Mydland was listening to the breakfast-time radio news from Oslo in his one-room apartment, which had a view right across the valley. From his window he could see the outline of the Plata Berget and the bright lights at the entrance to the mine workings on the steep mountainside below. He listened attentively to the broadcast because he was expecting both a worsening of East–West tension and something more, something of extreme concern to himself and two other people in the settlement. Although the Russians were already exercising control over the re-broadcasting of Norwegian television programmes, they had not yet interfered with the long-range national radio transmissions.

'The Soviet attitude is considered unacceptable by the Government,' the announcer was saying. 'A further strong protest is being made to the Kremlin by our Ambassador in Moscow and Norway will now ask the United Nations Security Council for a resolution demanding Soviet withdrawal.'

Mydland stopped spooning up his cereal to concentrate better. The news so far confirmed an escalation of the crisis. What next?

'Sources in Svalbard report growing discontent in the mining community there. The Justice Ministry appeals to all citizens to remain calm. Nothing should be done to prejudice negotiation. The Ministry asks us to repeat this advice. Nothing should be done to prejudice negotiation.'

Mydland felt his breath quicken in spite of himself. He reached over and switched off the big transistor set, then pushed aside the cereal bowl. He no longer felt like eating. That repeated sentence had been a prearranged signal to him and the two others who constituted an underground intelligence cell for the Norwegian Army in Svalbard. The others were a mine foreman called Haakon and a woman called Annie who worked in the local bank. She was their radio operator. The three never met all together, indeed only rarely met at all. The cell had last been activated during the dispute with the Russians over the Barentsburg helicopter base in 1978 and its personnel had changed completely since then, as members left Longyearbyen in the course of their normal civilian jobs. Tonight, in accordance with the signal, they would call the Defence Command for North Norway at Bodo in northern Norway on a miniaturized agent radio which could transmit a complete coded message in a condensed transmission of less than a second. It wasn't completely undetectable, but only a very sophisticated monitoring system would be able to pick it up. And it would still be in code.

Things may be worse than I thought, Mydland murmured to himself. Although the Treaty forbade the use of Svalbard for warlike purposes, he knew that the Army had contingency plans for a Russian invasion, and as a captain in a special branch of the Army reserve, the *skijeger*, he had been a natural choice for recruitment to operate here during an emergency.

Most countries have their special forces, small groups of men trained to act fast and decisively in unorthodox ways. Among them were Britain's highly competent if over-publicized Special Air Service, the Delta Force to which Lieutenant Colonel Tom Peterson had belonged in the United States, and the Soviet Union's Spetsnaz units, controlled by the KGB. Even the largest nation can only produce handfuls of men with the intelligence, courage and physical fitness required for special operations, so it was unsurprising that Norway's *skijeger* and *fallschirmjeger* – parachutists – numbered only some 120, of whom a third were regular soldiers. The majority were reservists, like Mydland, and had to keep quiet about this spare-time activity. As far as his friends were concerned, Paul Mydland had served, before he came to Svalbard, in the highly trained Home Guard, who kept their weapons at home and in time of war would go straight to preplanned defensive tasks. A neutral country until 1940, Norway had learned a lesson from being unprepared for the Nazi invasion.

When he had notified the Army of his new job in Svalbard, Mydland had been called to the headquarters at Reitan, a few miles inland from the port and county town of Bodo in the Arctic Circle. He had been given lunch by a major general, a tall bony man with an intelligent face and high forehead, who reminded Mydland of a university lecturer.

'No one is forcing you to volunteer, Paul,' the general had emphasized. 'This is outside the scope of any normal reserve liability. But we need people in place *before* anything happens. Which, pray God, it never will.'

'If we're not in uniform, surely they could call us spies?'

'Legally you will be Norwegians on Norwegian soil. You will have the right to operate wireless telegraphy equipment under Article Four of the Treaty. Not that anyone will be respecting the Treaty in the situation we're preparing for.' He had laughed. 'It will be possible to cache a uniform for use in extreme necessity. Anyhow, possession of a rifle comes naturally up there. No one in their senses leaves a settlement without one.'

Paul Mydland had agreed to accept the responsibility. He had been willing, ironically, because the German invasion of the Second World War was as real to him as it still was to many Russians. Norwegians had last been deprived of free speech and liberty by the Nazis. The resistance movement then had included his own uncle, who had been so tortured by the Gestapo at its headquarters in Tromso that he had leaped from an upstairs window when a guard was momentarily inattentive. It was assumed he had preferred death to the risk of betraying his colleagues. Although Paul had been born long after the war ended, his mother's memories had made the Nazi occupation as real to him as if he had actually suffered it. He was certain that a Soviet occupation would be no better, probably worse.

A year after Mydland arrived in Svalbard he had taken over the leadership of the cell from a man who was leaving. Locally he was not known as anything other than a clean-shaven, presentable young executive, with a reputation for not minding getting his hands dirty at work and for being a first-class cross-country skier at weekends. Like Annie, the woman in the network, he had the brown hair and blue eyes common to many Norwegians and was fond of joking that the only typical Scandinavian blondes one ever saw in Norway were air hostesses, and they were Swedes or Danes.

Paul liked Annie. If the security dictates had not kept them apart, he would have started dating her. As he washed up his breakfast plates and pulled on his parka before going to work he reflected that he would be glad to see her this evening, even though it would only be for the few minutes needed to call Bodo on the radio.

He finished at four that afternoon, left the mining offices and walked briskly up the road. Just past the two-storeyed, reddish brown building of the shop stood the bank, a small wooden cabin with the sign TROMS SPARBANK on its roof. Its windows were dark, as he had expected. He entered the brightly lit supermarket, but she wasn't there, so he went on to the bachelor apartment block where she lived.

Annie's apartment was tiny. A rectangular room with a window at the far end, a sofa bed to one side, a table. Near the door was a curtained-off recess into which she immediately retreated to make coffee. The recess contained a washbasin with a mirror above, a cooker and small fridge. Bachelor quarters in Longyearbyen were hardly luxurious, whether for men or the few single women.

'You've made it very nice,' he commented. She had indeed done her best with ornaments and colourful materials.

'Well, I am here for a year,' she called back. 'I hate living in a slum.'

She emerged, gave him the mug and a piece of cake, then came to the point.

'I heard the broadcast so I was expecting you.'

Without speaking he handed her the draft signal he had already encoded with a one-time pad. She went to a cupboard, hefted down a small suitcase from the top shelf and opened it. Inside, buried in underclothing was the radio.

'We ought to find a better hiding place now,' he remarked as she tapped out the morse letters.

'I'd rather not,' she said when she had finished. 'Everyone knows what everyone else does here. They're the world's worst gossips. If we start burying the set up the valley, or concealing it under the building, we'll be spotted at once.' She extended the aerial near the window and pressed the transmit button, then settled down to wait for an acknowledgment as calmly as if she had been telephoning her mother.

The operator at Bodo was obviously both alert and ready. The reply came in two minutes. Mydland decoded it and again showed her the pad in silence.

'Well,' she remarked after a few moments' thought. 'If they want information about the airfield, Frederik Folvik is the obvious person. He's very bitter about the takeover. I think he'd help us.'

The fifteen ambassadors in the United Nations Security Council sat round the circle of tables in their debating chamber. On the desk in front of them stood the name plaques of their countries, the United Kingdom's representative on the right of the United States' and close to him the USSR's. Behind them, their aides sat attentively listening to the translations of speeches through earphones: the Council worked in five official languages.

The draft resolution before the Council was straightforward: to condemn the Soviet Union's violation of Svalbard's territorial integrity and to demand an emergency special session of the General Assembly to consider it.

For Norway, whose ambassador was attending as an observer, the voting would be crucial. Long a contributor to UN peace-keeping forces in such diverse trouble spots as the Congo, Cyprus and the Lebanon, Norway had naturally first appealed to the UN. If the Security Council passed this resolution, Norway could immediately request that the United Nations send a neutral force to Svalbard, replacing both her own police and the newly arrived Russian Border Troops. The Soviet takeover would then be checkmated. The Soviet Ambassador could of course veto the resolution, as could any other permanent Security Council member, but the opinion of the United States and Britain was that the Russians would not want to use the veto at this stage. Put crudely, it would look bad internationally.

In consequence, the past two days had seen intense political lobbying in the corridors and offices of the thirty-nine-storey UN tower on New York's East River. The Secretary-General had been asked by Norway to put a resolution himself. Under Article Ninety-nine of the Charter he could bring to the Security Council's attention any matter which in his opinion threatened international peace. However, the Secretary-General felt it would be prejudicial to his future influence to lay a charge against one of the five permanent members. He was a cautious man.

Great Britain had agreed to initiate the resolution. Even so, as in any other international organization, the ambassadors had to bargain for support. China, also a permanent member, had threatened to abstain if the resolution was not tougher. The Chinese lost no opportunity to belabour the Russians. The French took a more conciliatory line, bearing in mind important trade agreements with the Soviet Union.

The Secretary-General opened the meeting with deep misgivings. For all the formal language of the resolutions, the deliberations of the Security Council often sank to the level of a badly disciplined school debating society. Some Arab and African nations were noted for their intemperance. This time it was the Chinese who let fly first.

'This interference in the Arctic, this seizing of an airfield, is a most flagrant violation of the Declaration on the Strengthening of International Security – '

'Point of order,' interrupted the burly Soviet delegate. 'Point of order. Norway agreed to this mutual defence arrangement in 1945.'

'That is not so,' the Norwegian observer interrupted. 'Our parliament rejected it.'

'We are not interested in the quarrels of the Great Power blocs,' shouted the Tanzanian, a smooth-faced rotund African in a Chinese-style cotton tunic. 'We should be condemning the racist South African regime. Only yesterday apartheid mercenaries . . .'

As the Secretary-General raised his hands, protesting against irrelevancies, the Russian smiled. He had laid careful plans for this meeting. Two African countries currently represented were so in debt to the Soviet Union for military aid that they would never support the motion. Tanzania was less in debt, but the delegate had readily agreed, over lunch in the delegates' dining room yesterday, that the Arctic was of no importance to the true freedom struggle taking place in southern Africa: fortunately he had been educated at Moscow University.

'The reason for putting this resolution before the Council,' the British Ambassador insisted, 'is that it is the duty of the United Nations to uphold the Svalbard Treaty, just as it upholds the Trusteeship responsibilities of the former League of Nations.'

He meant this as a sop to the Tanzanian, but it failed. All that ensued was a further outburst.

An hour later the Secretary-General brought the issue to the vote. A minimum of nine votes in favour were needed for a decision. The Tanzanian sat with his arms firmly crossed. The Angolan solemnly shook his head: Russian advisers and Cuban troops had supported his government for years. The Vietnamese, happy to offend China on a matter of no Far Eastern significance, also voted against. Two Arab representatives abstained.

The vote hung in the balance. The United States Ambassador, grey-haired, with heavy-rimmed spectacles balanced on a long nose, looked cautiously at the Ambassador of Salvador. An unfortunate quarrel had blown up with Salvador recently over a politician caught smuggling heroin into New York, whose claim to diplomatic immunity had been rejected. For a long moment the Salvadorean hesitated, then he too abstained. 'I do not see this as a threat to world peace,' he muttered.

'The resolution is defeated,' said the Secretary-General.

The Norwegian observer rose and left without a word. He was disgusted. The UN had failed one of its most loyal and consistent supporters. The one obvious way of averting an East–West confrontation over Svalbard by diplomatic means had been lost. His country must now turn to its NATO allies for support and, if necessary, action.

Tom Peterson's return to Norfolk was brief. The morning after his arrival home, Admiral King summoned him, listened to a summary of the intelligence he had gained on Svalbard, then told him that the Joint Chiefs of Staff in Washington had agreed a contingency plan.

'The satellite photographs and the information the Norwegians are passing us both suggest a radar site on the mountain above Longyearbyen's airfield,' King explained crisply. 'The Russians have had a survey team up there. Now, as I warned you, one way of dealing with such a radar would be to go in and destroy it clandestinely. If, that is, diplomatic pressure fails to stop them, and I for one reckon diplomacy will fail.' King almost snorted at the memory of the farcical Security Council meeting. 'So I'm designating you as Project Officer to train a raiding team from the Delta Force.'

'That's great, sir.' Peterson grinned. This was exactly what he had hoped, save in one respect. 'You wouldn't let me lead it in?'

'I'm sorry, Colonel. I'd prefer to have you control the operation. Maybe from Norway, maybe from a ship. That's your assignment, right? You'll be directly responsible to myself and a committee.'

'Will the British take part, sir?' There was no sense in arguing with the Admiral so Peterson went straight to the obvious query.

'We're hoping they will. More than them, we need the Norwegians, it's their damn territory.'

'I'm not too happy about the trawler those British commandos converted. Looked slow and vulnerable. Anyway it couldn't operate until the ice clears in June.'

'We'll see.' King temporized. 'With Norwegians aboard it would be pretty well camouflaged. Everything depends on the Russians, too, doesn't it? They might not build the radar until the snow melts. Right now, what I want is for you to start training a small group for a long approach march through terrain like Svalbard with an attack and a demolition job at the end. Precisely how you're inserted can be decided later, but you can assume a helicopter would lift you out.'

'I understand, sir.' Peterson had already decided where to train. 'I guess I'll take them to the High Sierras.'

'You're the expert. If you have any problems, come back to me.' King dismissed the subject for the moment. 'I hear your wife's expecting a baby. Congratulations.'

'Thank you, sir.' Peterson stayed noncommittal, wondering how Nancy would like his going off again. 'We're pretty excited,' he said. 'What with it being our first.'

King eyed him quizzically, guessing what was passing through his mind. 'Tell her from me you'll be home for the great event. The combined opinion of both Langley and Fort Meade is that the Russians have planned this well in advance. If we don't withdraw our ships from the eastern Atlantic, and, believe me, we shall do no such thing, they'll go ahead fast with whatever they intend because they'll want to catch us on the hop.' He shook his head angrily wishing that either the CIA at Langley or the National Security Agency had been able to provide prior warning of the takeover. 'As they already have done. So I'd like your team ready to go in two weeks. Thank you, Colonel.'

That afternoon as he drove home to Virginia Beach, Tom Peterson determinedly turned his mind to how Nancy would react. Hell, he wasn't unique. Other men's jobs took them away from home, travelling salesmen, airline pilots, dozens of professions. She was going to have to learn to live with it, though he had no intention of saying so that bluntly.

To his surprise, when he opened the door she ran out from the living room looking radiantly fresh, her dark hair shining, and threw her arms round him.

'Welcome home,' she said and kissed him.

'Hey, what a beautiful surprise.' He stood back a moment and gazed at her. 'You look like a million dollars. How come you escaped early?'

'No class this afternoon. The hell-raisers had a football game. And how come you're back so soon, for that matter?'

'Oh, the Admiral let me off the hook.' He knew it would be a mistake to explain yet. She was in such a happy mood that bad news would have to wait.

'Well, don't let's talk about him.' She sensed her husband's reticence. 'We've an hour or two of a lovely day left. Let's drive out some place. I feel as if I haven't seen you for years.'

'A week and a half!' he laughed. 'Anyone would think we haven't been married long. Let's go then. You're right, we should do something special.'

They drove out into the true countryside, with the sun low over the winter landscape and the air crisp and clean, stopping later for a drink and then a meal at a small restaurant. It was one of the most perfect evenings they had ever spent together. When they reached home again, she put on a soft music tape, fetched him a scotch on the rocks, nestled against him on the sofa and remarked almost casually, though not quite casually enough:

'So where are you off to next, darling?'

'Next? How in hell did you know?'

'It was written all over your face when you got back. So, I'm ready.' There was more than a shade of tension in her voice. 'Or is it classified?'

'No, it isn't. I'm going to the cold-weather training area round Bridgeport in the High Sierras.' He knew this would provoke an explosion and it did.

'California! Skiing! You're a bastard, Tom Peterson!'

'But, sweetheart, if you had a fall skiing you could lose

the baby. I'd like nothing better than to have you at Bridgeport. Not that you'd see me often. We'll all be sleeping rough.'

'It's all right, I do understand.' She held his hands tenderly. 'Who'll you be with?'

'A few of my old unit. I have to help train them.' He squeezed her gently and kissed her. 'That's rather more than I ought to say.'

She shifted round on the sofa and looked concernedly straight in his eyes. 'Tom, when this is over, whatever it is, would you consider leaving the Marine Corps, starting another career? I don't like thinking of you being in danger.'

'Sweetheart, I won't be. I tell you I won't, not this time.'

The reassurance hardly helped because she caught the regret in his voice.

'Will you think about it, darling?' she persisted.

'If you want me to, yes, I will.' It was the most honest answer he could give when the Marine Corps had been his whole life until he met her. 'Now – ' he deliberately kept his tone serious ' – since I have to go away for a while, can I ask you something, Mrs Peterson? You can't ski safely because of the baby, but can you make love?'

'Yes,' she said, hugging him. 'I most certainly can!'

The architect of the Delta Force was Colonel Charles Beckwith, the man who led President Carter's abortive attempt to rescue the American hostages from Teheran in 1979. Although he hadn't taken part, Tom Peterson had served as one of the so-called Charlie's Angels and always reckoned the Iran raid had political and administrative disabilities imposed on it which the most skilled leader could not have overcome. President Carter had postponed it nineteen times. The subject was much in his thoughts as he flew down to Fort Bragg in North Carolina to activate Admiral King's plan. He wanted to be damn certain his own side didn't wreck the Delta Force's chances again.

If Fort Bragg were superimposed on England, its training areas and barracks would stretch from Aldershot to London. It was a huge stretch of what the Army liked to call 'real estate', the home of two US airborne divisions as well as the Special Forces, the Green Berets.

The Delta Force had its own compound within Fort Bragg, where it selected the recruits for its two squadrons, a term borrowed from the British Special Air Service, which had helped with advice when the Force was formed in 1977. Every one of the few hundred troopers who wore its black beret had been through almost savage tests of physical stamina and mental endurance. All the officers were either rangers or airborne trained, often both. Peterson was one of the very few Marines who had joined the Force, partly because the Marine Corps had its own élite Force Reconnaissance teams to attract the ambitious.

The major general who coordinated both the Delta Force and the Green Berets welcomed him to a sparsely furnished, businesslike office and came straight to the point. His close-cropped hair emphasized a bullet head and he spoke with a clipped accent.

'We've assembled a group for you, Colonel.' He handed Peterson a list. 'Captain Howard Smith will be in command. Most of the others come from the mountain troop. We reckon he'll basically need a second in command, a medic who's done the dog lab course, a demolition expert, a communications man, and a Norwegian linguist. Will there be Norwegians taking part?'

'We hope so, sir.'

The general caught the doubt in Peterson's expression. 'You don't like that, huh? Too damn insecure, right? Well, I guess we have to live with these NATO allies.'

Fifteen minutes later Peterson was being introduced to Howard Smith, the officer on whom the success of the raid would ultimately depend.

Howard Smith was one of those archetypal East Coast college men whose regular good looks reminded one of the late President Kennedy. Clean-jawed, blue-eyed, open-faced, his green fatigues impeccably pressed. His military record, judging from the notes the general had provided, was equally impeccable and about as different from Peterson's own as it could have been, starting with the military school at Valley Forge and a congressman's recommendation, which had secured him a place at West Point. It flashed through Peterson's mind that Smith had been chosen because this might be an international operation, but he as rapidly dismissed the idea. The Delta Force set too high standards for that.

'Glad to meet you, sir,' said Smith conventionally as they shook hands. 'Seems a pretty exciting project.'

'A pretty tough one.'

'Tougher the better.' The incipient lines crinkled around Smith's eyes as he smiled.

No, thought Peterson, I can't exactly say why I wouldn't trust this guy the whole way. Too much like a movie hero, maybe. He wished he were allowed to select the team, instead of having that done for him. 'I guess we should get on with the briefing,' he said to the general.

Security in the Delta Force compound was tight at all times. The troopers were careful never to be named, let alone photographed, outside. They had it ground into them that publicity of the kind the Green Berets used to enjoy was not merely undesirable but downright dangerous. On this occasion there was a further identity check before the men filed into the briefing room to sit on tiered seats while Peterson began explaining the mission with the aid of photographs and viewgraphs brought from Norfolk.

'The code name allocated this mission is Virginia Ridge,' he announced. 'And I don't need to tell you that name itself is top secret. There's never going to be public knowledge of Virginia Ridge either before or after or, God willing, in thirty years' time.' If we can keep clear of the goddamn Freedom of Information Act, he thought. 'It's going to be one of the most sensitive operations the United States has been involved in since World War Two and we don't want any snafu with it.'

With this introduction, he launched into an illustrated description of Svalbard, of the strategic significance of the archipelago, and of what the Russians were believed to be constructing there.

'From the intelligence interpretations we're getting,' he said, pressing a remote-control button which flashed another picture on a screen, 'the radar's going to look something like this.' The drawing board showed a squat building, with a huge latticed radar scanner on a mast a few yards from it. The artist had sketched in snow and the plateau of Plata Berget. 'That building is our objective, when it's finished, that is, which we guess will be in around a month. By that time it'll be coming into spring there. There's no darkness and damn little cover.' He continued outlining the problems. 'The approach has to be clandestine. But when we've beaten hell out of that – ' He checked himself, remembering that he was not taking part. 'When *you've* beaten hell out of it, you'll be lifted out.'

When he had finished, a variety of questions ensued about the technicalities of weapons and radio. However, there was one point which was not raised and so he put it himself.

'You may have wondered why the CIA aren't carrying out this assignment?'

There was a murmured 'Because they'd foul it up' from one of the men, and a few laughs.

'No, that's not the reason. It is – ' Peterson looked them over as if awarding an honour, 'because the CIA lack operatives with the mountain-warfare skills you possess. OK? But there's a catch. I told you this was top secret. I meant it. Your families aren't to know of this either. It is a clandestine, repeat, clandestine operation.'

He paused to let this sink in and saw, painfully, in his mind's eye the pictures in *Time* and *Newsweek* of the burned crewmen on the Iran raid, their incinerated corpses lying in the desert by the wrecked Hercules plane. Would these men's bodies be put on display if the raid failed? Admiral King thought not. The Russians themselves had casualties from Angola whom the South Africans held, and dead from Afghanistan. The West could publish pictures of them if it wanted. Only bastards like the Iranians broke the unformulated rules.

'Be clear about one thing,' he said, watching the intent young faces in front of him. 'In theory there could be public outcry if you get killed. But there isn't going to be, because we and the Soviets are both likely to keep it quiet. Nor, if you succeed, are you going to be heroes except maybe here.' He automatically slipped into repeating the phrase from his platoon commander days. 'Be clear about that. It's a dangerous mission, but there's no funeral at Arlington if you fail and no presidential reception if you win. Any one of you who wishes to pull out at this stage can do so and no dishonour will attach to him. None whatever.'

Peterson surveyed his small audience, man by man. No one raised a hand or stirred. He turned to the major general. 'Thank you, sir. That's all I have to say.'

He hoped he hadn't been overly dramatic and sounded like a cheap movie hero himself. But only endurance and skill would enable so potentially hazardous an operation to succeed and he was going to drive these men extremely hard in preparing for it.

4

The Sea King helicopter scudded low across a grey and angry sea, the noise of the two jet engines a subdued thunder. By twisting himself round, the crew chief could see through one of the small windows. Outside it was snowing, the flakes clinging to the Perspex, dimming a visibility already hemmed in by the all-enveloping cloud, through which a faint yellow halo of sun occasionally appeared. As they headed farther north across the Barents Sea even this pale light would disappear. The Sea King, keeping a bare 50 feet above the cold waters, to escape detection, would reach Svalbard in the darkness of midnight and a temperature of around minus 27 centigrade.

Five men were strapped into the Sea King's canvas seats but only the crew chief wore a flying helmet and could talk to the pilots through his microphone. He remained alert. The lieutenant and the other *skijeger* sat hunched in the fatalistic, time-killing half-sleep which men waiting to go into action adopt. The noise made conversation difficult and in any case there was little to say. The briefings at the Defence Command headquarters were long over, committed to memory and map markings. The Sea King had refuelled briefly at Hammerfest for the five-hour flight and already mainland Norway seemed strangely remote. So the men slumped in their own private thoughts, or dozed, while out of their sight in the cockpit the pilots scanned the radar for the echoes of ships, the luminous dots on the round screen which would warn them to alter course. This mission was completely clandestine.

The lieutenant glanced back at the equipment tied down in the rear: demolition explosives, food packs, white camouflage netting, tent sheets and guns.

A sudden clattering of the helicopter's rotor blades shook the lieutenant back into the present. The Sea King was banking sharply. The pilot must have picked up a ship on the radar. He peered out of the window. There seemed to be a line of breakers in the distance. He consulted his watch, then understood. They were halfway, near Bear Island, and the line was the edge of the pack ice. In early summer this arctic barrier would retreat until it lapped the west coast of Svalbard and left the entrance of Longyearbyen clear. The polar bears went with the ice. If they missed their passage on the floes and were deprived of seals to hunt, they became dangerous to men. He reckoned the old-time trappers earned their money.

The thought brought him to his own mission. The Sea King was going to land briefly and unload near an abandoned trappers' hut on an uninhabited part of the Van Mijenfjord, 40 kilometres across the mountains from Longyearbyen. On the map it was 40 kilometres: 90 by the route the men meeting them would have to take back. He rehearsed the plan in his mind.

Outside it became dark. The crew chief signalled to them, shouting above the noise and sticking his thumb up confidently. The men checked their thick clothing, preparing themselves. Ahead of them the pilot saw a bright light pulse three times, then vanish.

The Sea King began to shudder as it hovered. The crew chief hauled back the sliding door, involuntarily gasping. It was like opening a giant ice chamber, the wind blasting in. Talking on the intercom, he helped the pilot inch down into the snow. The helicopter settled, the vibration ceased and the men leapt out to form a defensive arc around the machine. As the engines were shut down the rotor slowed and stopped, though the Sea King's machinery still made small whirring and hissing noises, as if refusing to die.

For a few moments the lieutenant stood still, taking stock, as the transition from the warmth of the helicopter hit him fully. The silence of the landscape was almost tangible. The ghosts of mountains rimmed the horizon, a single star shining weakly above. A paleness to the south suggested the coming daylight. His breath was a cloud of steam. This was where all the assumptions of civilization died, where mere survival was a struggle. Already the cold was starting to anaesthetize his cheeks under the fur-lined hood of his parka. He looked around carefully, fearing an ambush despite the light signal. The trappers' hut was a rough-hewn shape 110 metres away, but where was Paul Mydland?

A low whistle sounded from the direction of the hut. The lieutenant flashed his torch momentarily three times, shielding it with his glove. A pair of dark figures detached themselves from the shadow of the hut and glided across the snow. One was Mydland in a white parka, the other Haakon Sommervik, the mine foreman, a bigger man with a spade-shaped beard.

'The cache is ready,' Mydland said. 'You can unload.' He pointed to where a snow-covered outcrop of rock made a faint outline like a whale. 'It's in a north-facing drift. It'll last into summer.'

'Took us two bloody hours to dig,' Haakon added with feeling. 'We're bloody frozen ourselves.' He had the vocabulary of a typical sergeant.

Using a simple flat-bottomed sledge called a pulk the *skijeger* towed the stores to the snowhole. When everything was stacked in the cavity, they closed it with snow. Mydland confirmed the compass bearings to the hut which he had taken earlier, and they retreated slowly to the helicopter, brushing the surface to fill in their tracks as they went. A few hours' wind or a light snowfall would obliterate the marks completely.

'Best of luck, sir.' The lieutenant saluted, then spontaneously gripped Mydland's gloved hand. The *skijeger* scrambled back up into the helicopter, the engines began to whine, the rotor turning faster. Mydland and Haakon crouched to one side shielding their faces, and the Sea King lurched up into the air, the downwash driving a blizzard of snow around them. Then it tilted forward and whirled away, back along the dim white surface of the fjord.

The two men filled in the deep indentations left by the weight of the Sea King as best they could, then returned to the hut. If the helicopter had been heard but not seen, the immediate presumption of any Russians would have been that it was one of their own.

The hut was fashioned of heavy timbers, shipped in many years ago for the purpose, since no trees grew as far north as this. It had two small rooms and had been used recently, though precisely when it was impossible to tell: the climate preserved everything too well. An old kettle stood on a plank table, there were ashes from a fire. Probably some expedition had been here last summer. Though too obvious to use for the cache, the hut was a bonus for one night.

Despite their physical fitness both men were exhausted. They had been alerted by Oslo a week before, after it became obvious that the Plata Berget was being surveyed and Folvik had reported recognizable parts of a large radar installation being unloaded one night when he was on duty. Yesterday, ostensibly on mining survey business, they had flown in a light plane to the small mining settlement of Sveagruva on the Van Mijenfjord, in the process of being re-established when the crisis arose. They had slipped out of Svea twenty-four hours later on a borrowed Skidoo, again on the pretext of surveying, sped along the frozen surface of the fjord and sheltered in the abandoned hut, one of many dotted around the archipelago, relics of days when hunting bear, seal and arctic fox were the main human activities here.

During the night the wind increased. When Mydland woke, he imagined snow piling against the side of the hut. He partially unzipped his sleeping bag, sat up and hastily pulled on a thick sweater, then finished dressing. In the Arctic you clothed yourself on the layer principle, each layer trapping air warmed by the body. He wriggled his feet into his boots and went to the door. Outside the scene had changed. All he could see in the early daylight was misty cloud. Windblown snow had already filled in the tracks they had made on arrival. He gritted his teeth. The Skidoo's progress would be seriously hampered. He backed into the hut again and shook the Sergeant awake.

'Haakon. We need an earlier start. The weather's closed in.' The ex-sergeant was known to everyone by his christian name, perhaps because Sommervik was long-winded to pronounce.

Haakon grunted, shook his head bemusedly, then was awake and squirmed out of his bag. 'Back to civilization, is it? Thank God for that.'

The Skidoo was a bright red and yellow painted snow scooter, propelled by tracks at the rear, with small steerable skis at the front and a long black seat on which two men could ride. It was a standard means of personal transport in Svalbard and could tow a small sledge for freight.

An hour later the fibreglass sled was laden with a bright orange civilian tent, cross-country skis, their two .375 calibre hunting rifles, a can of fuel, sleeping bags and a battered suitcase apparently stuffed with clothes, in which a radio for emergencies was concealed. Both men had tugged covers on their boots to stop sweat freezing in the leather.

'You'll ride shotgun, then,' said Mydland cheerfully.

Haakon nodded. They pulled goggles over their eyes, wrapped scarves around their faces, mounted the Skidoo and set off cautiously. Driving the Skidoo was like steering a motorcycle, except that the tracks kept it upright. The problem was vision. At first, following the frozen shore of the fjord, Mydland could see enough. But their route went up a wide valley called the Reindalen, off which another led to a 1300-foot-high pass. The invisible sunrise had hardly lightened the landscape enough for them to see the steep sides of the valley, when their course forced them up into the misty cloud. The visibility fell to 50 yards, to 20, to 15. All definition of the snow's contours vanished in the haze, hollows becoming indistinguishable from rises. The Skidoo repeatedly ran into humps and they had to pull it out.

For an hour Mydland persevered, glancing frequently at the compass he had mounted on the handlebars, the engine grinding tinnily in bottom gear. Then he stopped and turned to his companion.

Haakon's moustache, exposed between his goggles and his scarf, was studded with beads of ice. He pulled the scarf down. 'Uses a hell of a lot of fuel in first,' he said. 'How far do you reckon we are?'

Mydland fumbled with gloved hands to unzip a pocket and clumsily extracted an altimeter. 'Thirty-one metres above sea level,' he announced, then pulled out a map. 'We could be 20 kilometres up the Reindalen. If we're in the middle. It's not only fuel, it's finding the pass. We'll have to wait for this to clear. Let's pitch the tent.' The next priority was shelter.

'Out with the shovel, then.' Haakon swung himself off the seat. They would dig down to make a floor, and at the same time cut blocks of snow as a barrier against the wind. Covering the Skidoo, they set to work. Both the cold and the heavy clothing turned normal activity into heavy labour and they wanted to avoid sweating, since the moisture would eventually freeze. So they moved slowly.

It was while they were fixing the ropes that Haakon sensed rather than heard the intruder.

'Listen,' he hissed. 'We're not alone.'

Mydland stiffened. They both stood still. From nearby came a low, wheezing noise, like deep breathing. But they could see nothing. The noise continued, sounding closer.

'It's a bear,' said Haakon, quietly.

Mydland stumbled through the snow to the sled, heaved up the rifles and hastily passed one to Haakon. Both men slammed cartridges into the breech and waited again, looking round uncertainly. The sniffing continued, the snow muffling the sound. Mydland had never seen a polar bear outside a zoo, but he needed no reminding that a full-grown one weighed half a ton and could crush the Skidoo, quite apart from killing them.

'Left,' whispered Haakon.

Taking shape in the fog was a black-nosed head, long-jawed, with small round ears laid back. The head swayed slowly from side to side, the huge body a creamy blur behind it. The bear stopped, one massive furry leg ahead of the other, its nostrils steaming as it exhaled.

'Hold your fire,' Mydland ordered. 'It may go off.' He stayed motionless, holding the butt of the heavy rifle in his shoulder, warnings about bears running through his mind. The animal could be merely curious. It could also have a mate somewhere out of sight. They had only four rounds each in their magazines. He watched the bear's eyes, small and dark above the long muzzle. It shifted a step, raising its head and making a different sound, as though smacking its lips, its teeth showing. Then without warning it bounded towards them.

Mydland fired, aiming for the shoulder. The bear reared up momentarily, snarling, then lunged forward again. It was incredibly agile for its bulk. As he worked the rifle bolt, reloading, Haakon's gun half deafened him. The bear was hardly 5 metres off. He dropped on to one knee to steady himself and quickly fired again. The bear shambled on, staggered and suddenly collapsed on its side.

Haakon stepped up to its heaving body, thrust his rifle against its ear and fired again. The bear shook and lay still.

'That was close,' Mydland said, trembling slightly in spite of himself. He stood, head cocked, listening, his rifle cradled in his arm. Nothing stirred.

'Must have been a loner,' said Haakon.

Together they examined the body. One shot, probably Haakon's, had thumped into its spine behind the head, from which a coagulating trickle of blood stained the snow. There was another small hole in its shoulder.

'So,' remarked Haakon coolly. 'Now I suppose we have to explain to the Sysselmann why we've killed one of Svalbard's sacred bears.'

'By God, yes! We have to do exactly that.' Mydland looked at the Sergeant, an idea crystallizing. 'Have you ever skinned an animal?'

'You're joking! A brute this size, in a temperature of twenty-six below?'

'What else are we going to do for the next hours?'

'Personally, I'd rather sit in the tent and watch television.' Haakon grinned insubordinately. 'Couldn't you just buy Annie some perfume in the shop?' He was about to expand this hint at Mydland's having an eye for her but thought better of it. The Captain's mild manner was deceptive.

'Get out your knife. This pelt could be the most valuable you ever set eyes on.' Mydland reached for the long-bladed, bayonet-like knife hanging in a sheath on his own belt. The arctic knife was symbolic of the far north. It was the friend and standby of soldiers and hunters alike.

'Cut from the throat along the belly,' advised Haakon. 'Then inside the legs. That's how we skin deer.'

Together they heaved the bear's body over and began, pausing often to consider the next stage. It was not as messy as Mydland feared because the flesh began to freeze within seconds of being exposed. After two hours only the head defeated them. They left it intact, severed the neck and stuffed it with snow, then laboriously folded the frozen pelt. Finally they cleaned their bloodstained gloves and wiped the knives. The carcass lay in a curiously innocent heap on the reddened snow.

'Dropped off the back of a butcher's truck, I'd say,' said Haakon with satisfaction. 'Christ, what an animal!'

'Let's brew some coffee.' Mydland gave his attention to their surroundings again. The amorphous mist seemed to be ragged in places now and the visibility had definitely improved. He pulled back his sleeve to check his watch. It was nearing half past eleven.

'We may be lucky yet,' he said.

By the time they had melted snow in a pan on the spirit stove, made coffee and eaten some biscuits, sheltering in the tent, the sides of the valley had come into sight. The cloud was lifting rapidly. They packed up again and loaded the sled with the bearskin tied on top like a tarpaulin. Within half an hour of setting off, the peaks began to show, jagged and monstrous, rising to 3000 feet on either side. They found the side valley without difficulty and ground noisily up the ever narrowing cleft on the mountains, until they came through the pass and the long arm of the Adventdalen lay before them under a fading blue sky. Soon it would be dark again.

There was a road along the bottom of the Adventdalen, leading to a radio installation associated with the airfield. Some way off Mydland stopped and studied it through binoculars. Soldiers were on guard. 'Now for the Russian bears,' he said tautly. 'Let's hope they appreciate our Norwegian variety.'

The border policeman signalled them peremptorily to stop and spoke in Russian. Mydland firmly shook his head, smiling regret at not understanding. With another gesture, the man made it clear they must not proceed and went to the small building. He returned with another who said in heavily accented Norwegian, 'You must wait.' Then the bearskin caught his eye. Mydland smiled again, dismounted from the Skidoo and began to mime the attack by the bear and how they had shot it. The Russians both nodded appreciatively, felt the stiff pelt and examined the head. By the time a Norwegian policeman had arrived on another snow scooter a definite rapport had been established.

'They ask, what will you do with the skin?' interpreted

the Norwegian after he had examined Mydland's papers and explained them to the Russians.

'Take it to the Sysselmann's office as the law requires.'

The Russians looked disappointed, but grunted agreement. They waved the two men past without making further examination of the sled. The Norwegian followed them down. Entering the settlement, ablaze with lights since it was now dark, they stopped. The policeman drew level with them.

'Any objection if we report tomorrow?' asked Mydland. 'I want to tell my friends I'm here.'

The policeman shrugged his shoulders. 'I wouldn't leave it. They're tightening up every day. How did you get out of Svea?'

'No problem. We left there two days ago. Been making an official survey.'

He gave them a sharp look and accelerated off, dirty snow flying up from the scooter's tracks. Even in winter there was coal dust in Longyearbyen.

'So. Does the bear get the medal or doesn't he?' demanded Haakon. 'Which d'you reckon got us through?'

'I'm not sure, not sure at all. The sooner we hide this radio, the better. Let's find Annie.'

'Safer if only one of us goes,' Haakon grinned. 'I'll go over to the Store Messe for a meal. If there's any food left.'

They drove to the miners' accommodation block where Haakon lived and unloaded his possessions from the sled. They both looked at the huge dirty white bearskin.

'I'm not keeping that under my bed,' Haakon said. 'Once it unfreezes it'll need a place of its own.'

'We'll have to leave it on the sled until tomorrow,' Mydland decided. He accelerated off, parked near his own living quarters and then walked the remaining distance to Annie's, carrying the old suitcase.

She let him into her tiny apartment cautiously and then clasped his hand in welcome.

'Was it all right, Paul?'

'No problems,' he said, happily pulling off his parka and feeling the warmth of the room. 'At least no serious ones. God, it's good to be inside.'

'If I'd known what time you were coming, I'd have prepared food. We could go to the House.' She saw the look of doubt on his face. 'Well, the whole block will know you've come back and visited me by tomorrow. It would seem more natural if we went to dinner, surely? That wouldn't be so bad, would it? In fact I should be delighted.' She had a wide generous smile with a touch of whimsicality which went straight to Mydland's heart.

'I'm sure you're right,' he conceded happily. 'People would think it odd if I didn't take a pretty girl out.'

'Have a drink then?'

'A whisky would be good.' He settled himself down into a chair.

'So,' she asked, 'what was the not serious problem? Were you stopped?'

He told her about the polar bear and the Russian sentries.

'We are in a very curious situation here, aren't we?' she reflected. 'Quite unreal. The Russians don't imprison people, if they do stop them for some reason they take them immediately to the Sysselmann's office. They are making a great pretence that he still has control.' She made a gesture with her hands. 'How long will that last? Their police are obviously under very strict orders.'

'Do you think they get much help from the local communists?' Like most mining communities, Longyearbyen had its dedicated members of the Communist Party. 'The miners I know profess to be shocked at the takeover.'

'Others are less loyal.' Annie was worried. 'The Russians had good contacts here before they came. Although, it's a funny thing, there is a lot of resentment. The Longyearbyen communists want to be communists in their own way. They don't like taking Soviet orders. And then there's the living problem. All those border police have had to be housed.'

'True,' Mydland agreed with her. 'It wouldn't be hard to

promote unrest. However, the officer with the helicopter told me our main concern must be intelligence on the airfield. Do you think we could actually recruit Folvik?'

'He's talked to me freely enough. You could only try. I'm sure he wouldn't betray you.'

'You'd better fix a meeting.' Mydland hesitated. He had something else on his mind. 'Listen, Annie. I've brought back the radio, but wouldn't you rather I kept it?'

'What happens if the Russians arrest you? There must be greater danger for you than me. The worst would be that I get deported.' She shook her head vehemently, her long brown hair swishing on her shoulders. 'No. We each have our jobs, the radio is mine. Let's go and eat.'

Makarov looked up at the mountain from his office in the Longyearbyen control tower. For the first time in many days the layers of cloud which had lain like a mantle over the summit were breaking up into ragged grey wisps. He picked up the phone and called Stolypin.

'Today will be suitable, Comrade General.' He preserved strict formalities with his senior.

'So I see.' Stolypin never gave away a point if he could avoid it. 'We will go at midday then. Also send out two helicopters on reconnaissance. Search the whole area south of here. You know what to look for.'

'Yes, sir.' Makarov rang off and set about turning this vague direction into precise orders. He shared the General's view that the two Norwegians the border police had interviewed were probably subversives, but where had they been? The only way to scrutinize Spitsbergen's precipitous mountains and hidden valleys was from the air and low flying had been too dangerous recently. It had also made the General's personal survey of the proposed radar site on Plata Berget impracticable. Not that this worried Makarov unduly. All other preparations had been going with remarkable smoothness.

Unaffected by the cloud, thick-bodied Antonov transport planes had been flying in and unloading almost daily, disgorging their cargo during the brief night, when the airfield was kept closed and most Norwegian staff were out of the way. The big hangar alongside the control tower had been cleared of the snowploughs and tractors which it normally sheltered and the scene inside its high arched roof resembled the assembly line of a factory. Tiered storage racking, carefully divided into bays and numbered, held the minor components of the radar which would soon be erected on top of the mountain. Larger items, like the steel framing for the control centre and the wide span of the main radar scanner, rested on the floor. The cafeteria, situated on a gallery inside the hangar, had been turned into a canteen for the technicians, and the rooms formerly devoted to the use of passengers had become offices. The whole area was heavily guarded.

It was now the second week of April. On the 20th, punctual to the day, the sun would fail to set on Svalbard and the long summer of perpetual light would begin. But it would be June before the snow melted in the township and July before the mountains above were clear. The question Stolypin had to settle was when construction should start.

The General arrived at noon, thickly clad, his fur hat clamped firmly on his balding head. He was accompanied by an Air Defence Command colonel and young Yakushkin, the radar expert. They spent a few minutes in a specially prepared office reminding Stolypin of details of the plan.

'Our problem, Comrade General,' the colonel explained, 'is a combination of radar coverage and interceptor range.' He pressed the remote control of a videograph machine and a projection of the northern hemisphere seen from above the Pole flashed up on the screen.

'This is the farthest out that an American B-52 bomber can launch a cruise missile aimed at us,' he said, then pointed to an arc cutting the eastern shore of Greenland. 'And this, Comrade General, is the maximum radius of action of a Kola-based TU-28 interceptor: 2000 kilometres, but still 500 short of where the B-52 releases the missile.' To remind Stolypin he quickly showed a photograph of the big, swept-wing, supersonic TU-28. 'If we used this airfield,' he went on, 'even the MiG-25s could intercept the launch point.'

Stolypin grunted, shifting slightly in his armchair. He was a thin man and however well upholstered the seat, his backside rapidly began to itch. He had already spent all morning sitting down.

'What's all this short-term special pleading, eh?' he demanded. 'We all know you'd like Longyearbyen airfield. But first things first. We can't bring interceptor aircraft here without precipitating a far greater international crisis than we want. Radar is something different. It is defensive. It can be argued over at the International Court for years. Surely radar here will give you the extra time you need.' He gave the colonel a sour look. 'Even against an unarmed airliner.'

The colonel gulped at this open rebuke, which reminded him forcibly of a most embarrassing incident. In the spring of 1978 a Korean Airlines Boeing 707, whose crew had made a cumulative navigational error on a trans-polar flight, headed over the Kola Peninsula towards Moscow. Although the Air Defence Command possessed a huge force of aircraft, defensive missiles and early warning radar, the Boeing flew on for three hours into the heartland of the Soviet Union before it was intercepted and forced to land on a frozen lake. The affair reflected little credit on the decision-making capabilities of the PVO Strany, as the Air Defence Command's title was abbreviated in Russian.

'That incident underlines the need for improved early warning, Comrade General,' the colonel said defensively.

'Then get on with it and show us those plans.'

The colonel motioned to Yakushkin to take up the story.

'We shall build two systems, Comrade General.' Yakushkin might be a technological wizard, but he was nervous and he fingered his collar. 'One will be for long-range strategic early warning, the other for more local surveillance. The long-range masts, which we'll have to ship in, will be erected near the existing Norwegian radio station near the entrance to the Isfjord. They will be able to identify both aircraft and ballistic missiles as they leave North America. They will transmit data back to the PVO Strany at Arkangel via a satellite stationed over the Equator. The Honeywell 2000 will process that data.'

Stolypin permitted himself a patronizing smile. He knew all about the ways in which embargoed US technology was acquired through companies in Western Europe. He liked the irony of American planes being tracked with the aid of an American computer.

'Can you quantify the advantages this will give you?'

'It will mean, Comrade General, that a B-52 or a B-1 flying across the Hudson Strait at an altitude of 50,000 feet will be identified on the controller's video within ten seconds of the radar at Isfjord picking it up. That is allowing for delays.'

When I was a young officer, Makarov thought as he listened, I would have been amazed at this, amazed and excited. Here is our motherland, less than seven decades freed from the serfdom of the czars, achieving technological miracles. Then he remembered the Honeywell computer and the fact that the unbelievable speed of microwaves was much better exploited by the Americans.

'And if the aircraft do succeed in launching their missiles?' he asked.

'Then the surveillance radar on the Plata Berget mountain above Longyearbyen will pick them up. The two systems make an excellent combination.'

'In terms of interception time, what improvement will it give?' Stolypin demanded.

'The subsonic cruise missile would be on our video up to an hour and a half earlier; the ballistic missile about half a minute.'

Stolypin nodded. Intercontinental ballistic missiles, streaking 200 miles above the earth to plummet down and flatten cities, presented such split-second interception problems that he personally believed no world power would ever use them for fear of retaliation. However, the slow-moving cruise missile with its 'battlefield' nuclear warhead, designed to destroy individual companies of tanks or single airfields, was a different matter. It was to nullify the American use of cruise missiles that the peace campaign in Europe had been orchestrated against their deployment. It was to nullify their alternative launching across the wastes of the Arctic that he was here in Svalbard.

'A very clear explanation, Comrade Yakushkin,' he said. 'Now let us see the mountain.'

They went downstairs to the waiting MI-8 helicopter and took off, the pilot soaring upwards with giddy speed.

Makarov watched intrigued as the contours of Plata Berget were revealed. Part of the way up, where a bowl-shaped natural amphitheatre cut into the mountainside, there was a mine with a few buildings. Above this the snow-laden rock was almost vertical. Farther round from it, a V-shaped valley cleaved the whole way to the summit.

'They call that the Blomsterdalen,' Makarov explained over the headphones. 'Because of the wild flowers in summer. We could run the data cables up along it.'

Stolypin merely nodded. He was intent on the landscape: the wide fjord stretching out beyond the airfield, the distant peaks on the far side of the ice. You did not need to be an expert to appreciate what advantages this magnificent field of view would give the radar.

The helicopter swayed in the wind as it traversed the summit of the Plata Berget.

'Not entirely flat,' commented Makarov. 'But a tabletop compared to anywhere else.' Ahead of them stretched a slightly humped snowfield two and a half miles long and half a mile wide. Makarov directed the pilot to land and they came down gently into snow which was less deep than might have been expected. 'The wind blows much away,' he explained.

The cold blasted them as they climbed out. Fifteen hundred feet of altitude lowered the temperature and the wind added its chill factor. Stolypin stamped his feet and looked around. Behind them, to the south, a ridge reared up another 1500 feet.

'So,' he half shouted, 'is this the site? How about that?' He gestured with a gloved hand to the higher mountain.

'It only obstructs to the south.' Yakushkin pointed in the opposite direction. 'Towards America, it's perfect. The surveillance radar will identify cruise missiles or low flying aircraft 400 miles out.'

Stolypin spent a quarter of an hour considering the conditions. They would have to build a camp up here. If the weather closed in, the construction workers would be completely isolated. It was hard to imagine a more exposed site.

'When will we start construction?' asked Makarov.

'In summer, Comrade General, when the snow has gone. That would be best.'

'What isn't best in summer, Viktor Mikailovitch?' Stolypin stumped around in his heavy boots, making a final survey, then returned to the helicopter. 'I shall recommend starting as soon as sufficient stores are assembled below,' he said uncompromisingly. 'What did you get the Nevskii medal for, eh? Sitting on your backside?'

Makarov bit back a retort. It was exactly the kind of snide remark headquarters officers make and he did not respect Stolypin for having reached the top through assiduous staff work, rather than leadership in action. Makarov himself was the kind who led from the front and didn't mind getting his hands dirty. 'Ice and snow obviously make accurate construction harder,' he answered woodenly.

'How do you suppose we have developed Siberia? Not by waiting for warm weather every year, I can tell you. In any case, the ground is still frozen in summer, is it not?'

'Except for the very surface, yes,' Makarov admitted.

What scientists called the permafrost was indeed permanent and it started a foot down.

'Precisely. Believe me, comrade, if the radar is to be operational before next winter, you have no time to waste.'

Makarov agreed, keeping his thoughts to himself. Stolypin really was a bastard. 'We' have successfully developed Siberia, but 'you' carry the can for the future job. He decided to throw down a minor challenge. 'Do we expect any military interference, Comrade General?'

'From NATO?' Stolypin could not suppress a smile born of superior knowledge. 'I think not, comrade, I think not.'

The intelligence advisers had assured him that they would be forewarned of any NATO moves. Then he remembered those two Norwegians. They might spy but would Norway ever dare attempt military action? He doubted it. However, the essence of success on the staff is to prepare against every eventuality.

'There must be a defensive plan, of course,' he said briskly, his words fighting with a gust of icy wind, and his eyes watering as he again surveyed the area. Even a conscript private could see that this exposed plateau would be a death trap to any force attacking it by land. Surely such idiocy was highly improbable. 'The real battle will be political and fought far from here,' he said, turning back to the shelter of the helicopter, leaving Makarov and the others to follow.

On their return Makarov ordered food to be brought to his office and began formulating a simple defensive plan for the radar while the site was fresh in his mind. Basically the Plata Berget was oblong in shape and threequarters of its perimeter consisted of unscalable cliffs. Well, he admitted to himself, not unscalable by a mountaineer, but extremely exposed to view. Only from the mountain ridge to the south was there any kind of secure, unobserved approach. Between it and the radar lay a 3-kilometre wide stretch of flat land. Or rather, at this time of the year, snow. That, he decided, was where any attacker would be caught in the open. That, in the phrase they had taught him long ago at the officers' school, would be the killing ground.

Cloud drifted around the mountain, making the distant peaks only occasionally visible beyond. The lower slopes were dotted with fir trees, their shadows lying long across the snow, which was already thin in places. Mid-April was the official end of the skiing season at most High Sierra resorts and the training time left to the team was now limited. Not that this unfrequented wilderness of snow on the borders of California and Nevada was a resort. Far from it. For the last three days the team had been trekking a long cross-country route to a mountain with a 9000-foot ridge and a plateau below it. Few people ventured here and they could practise their final assault on the radar in privacy, though not firing live ammunition.

The ridge and the plateau were not the same as the Plata Berget above Longyearbyen, but they had one important aspect in common: the snowfield was exposed and the small hut erected on it to simulate the radar installation would be extremely hard to approach undetected. Peterson was inside the hut and the assault was imminent.

'Let me tell you, Clifford,' Peterson remarked to the officer standing at the window beside him, 'getting to this objective is some problem.'

'Why not airlift them straight onto it?' The other officer, Major Craig Clifford, was a tall, well-built thirty-two-year-old, whose parka and overwhite clothing made him seem more thickset. His jutting chin and broad cheekbones would predictably flesh out in later life to form one of those strong, dominating faces characteristic of tycoons both in fiction and life. His classmates at the officers' school had nicknamed him 'Four-Star Clifford' before he was even commissioned, and he spoke with the incisive brevity of a leader.

'A couple of Sea Knights could bring them,' Clifford

argued. 'Wham, bam, the whole damn thing would be over in minutes.'

Peterson glanced round. 'That's my instinct, Major. But there's just about the whole Soviet Northern Fleet in those waters. The choppers would be identified way out at sea and the Admiral reckons there'd be a reception committee waiting. Kind of a hot welcome too. So we go for the clandestine option.'

'And pray for bad weather?'

'Right. Needs to be worse than this too. Howard Smith's going to be forced into a frontal assault, though he's trying a new idea with a missile.'

'Hmm.' Clifford grunted, gazing out at the cloud filling up the valley below. 'How will Smith take the news?'

Peterson was also searching the landscape and he did not look round. 'Smith may not like it too much. But be clear, the British are attaching a commando element to us under a major. We can't have Smith outranked. Either he has to be promoted or replaced. He'll just have to accept that politics are involved.'

'I hope he sees it that way,' Clifford said shortly. Small and hand-picked the Delta Force might be, but it was not immune to professional pride and rivalry. If Peterson's debrief after this exercise was critical, then the announcement of a change in command might well be wrongly construed.

Both men fell silent, waiting for the attack, whose success they must judge without a live shot being fired. Behind them a signaller crouched over a small VHF radio set, the PRC-77 type which worked better in the mountains than the standard infantry wireless. The signaller had fixed it so that they could all hear. Ahead of them cloud drifted like skeins of white wool, occasionally parting to reveal the ridge.

'Two missiles fired. Target destroyed.' Corporal Trevinski's harsh voice came so loud over the radio that Peterson jumped. He and Clifford were the target. Where in hell was Trevinski? Far off to the left, at a good 2000 metres' range, he glimpsed two men kneeling in the snow and the angular shape of the missile launcher.

'Well,' he said to Clifford. 'They sure got in range. I guess this dump's officially on fire.' He pulled open the door and ran out, reaching for the M-16 carbine he had propped against the wall, with Clifford and the signaller following. The snow outside was windswept and crusted enough to stand on. He glanced all round.

'I have them,' Clifford shouted. 'Ten o'clock, 600 metres.'

Emerging from the cloud ahead and to the right were four men, skiing laboriously, weapons in hand.

From the left came Trevinski and the young lieutenant, Andrews, who both dropped flat at a few hundred metres and began simulating rifle fire, the thin crack of blank ammunition sounding even less genuine across the snow.

'Tell them it's all over,' Peterson ordered the signaller. 'Close on me.'

Five minutes later the whole team was standing in a semicircle around Peterson and Clifford, leaning on their ski sticks, the effort of three days' unrelenting march evident in their faces. Peterson nodded at them one by one, checking their appearance.

Basically they all wore standard arctic clothing, with cotton camouflage suits buttoned over it. These might be called 'overwhites', but after three days they were soiled and torn and only partly concealed their back packs and equipment. Their webbing war belts were masked with white tape in places, but some items remained dark, like the entrenching tools. All the men had added thermos flasks to their kit, since the issue aluminium water bottles invariably froze solid in deep cold. The flasks were taped to the web gear. None of the men had steel helmets, because in the snow there was less danger from flying shrapnel. Instead they wore knitted woollen caps under the lined hoods of their parkas. Their weapons were not standard. Trevinski, for example, had equipped himself with a modified M-16 with a telescopic sight. Most carried knives as well as guns. What mattered was that every man was a marksman with his chosen weapon.

Howard Smith, who had managed to shave as always, stood in the centre of the semicircle, flanked by Lieutenant Jack Andrews, like Smith, a former college football star. The youngest was Burckhardt, a lean farmer's son from Wisconsin who was the radioman; the corpsman was Johnson, with the first-aid box in his pack; the linguist – as much a key man as anyone – was Sergeant Neilson, who doubled as a demolitions expert to assist Trevinski, a handsome twenty-eight-year-old with thick black hair and a moustache to match. Off duty Trevinski pursued women with legendary determination and was the inevitable butt of jokes about his Polish origins.

Each man had two skills. Johnson, for instance, was a radio operator as well as a medic. Out in the snow they paired up to dig snow holes and sleep on the 'buddy-buddy' system, one cooking the hot meal of the day from the freeze-dried rations known as LRPs or 'lurps', while the other cleaned equipment or kept watch. To a conventional infantry or Marine officer the team would have seemed absurdly small: a normal Marine combat patrol would number forty men. But forty men make noise, leave trails and litter, need constant shepherding. The Delta Force team would operate with the minimum of orders and the maximum of concealment. If any group of men could get into Svalbard unobserved and blow up a radar, this was the group.

'Well,' Peterson said, surveying them. 'That was better than some cluster screw outfit might have done. But you'd need that main fire group closer if there was any kind of an organized defence.'

'Hell, Colonel, those missile rounds would've caused enough confusion, wouldn't they?' Howard Smith's eyes were red-rimmed, he had obviously driven himself hard and there was an aggressive defiance in his question.

'Be clear, Captain.' Peterson didn't believe in letting the usual Christian-name informality undermine discipline. Like most ex-sergeants who have risen to commissioned rank, he was wary. 'If that radar's as large a construction as we anticipate, then you have to get right in there and place demolition charges. The missile's a good idea, sure. But unless things change, it won't do the job by itself.' He turned to Trevinski. 'How about the missile, Sergeant? It's a lot of weight.'

'Humping it's an ass-kicker, sir,' Trevinski replied with feeling. 'OK, so it does break down into four loads and, yeah, it's a great weapon, sir. Question is whether it kills us before we get to use it.'

Peterson laughed appreciatively. In this outfit a corporal could speak his mind and Trevinski put it well. The TOW was a wire-guided anti-tank missile with a 3000-metre range. The guidance system meant the firer could control it all the way to the target. Howard Smith had conceived the idea of solving the approach problem by disabling the radar and causing confusion from a range far outside defensive small-arms fire. But the TOW system was heavy. True, on the march it could be split into its guidance set, traversing unit, tripod and launch tube, each one being carried by a different man. But there were still the missile rounds and it wasn't as though the team were travelling light in the first place. Apart from their personal weapons, they each had four days' rations, a tent sheet, bottles of fuel for the stoves, candles, batteries, a 15-metre length of line, eighty rounds of ammunition, snowshoes, washing kit, spare socks and a change of special mesh underwear. They would be humping 70 to 80 pounds without the missiles.

'We could use a pulk, sir,' Lieutenant Andrews suggested. The pulk was a lightweight Norwegian sled, roughly six feet long and one and a half wide. Like a child's plastic sled it had no runners. It was pulled along by ropes and that was the snag. Two men, maybe four, had to be harnessed to those ropes.

'Well,' Peterson did not want to put down Andrews too hard, since he was about to remove him from the team anyway, 'you all know how I feel about pulks. They're fine on a big patrol, but they cost you mobility and speed of reaction. Let's work on this question of the missile.'

'How about the Dragon?' Clifford suggested.

'Only 1000 metres' range, sir!' Smith cut in heatedly.

'Still keep us clear of rifle fire, sir,' Trevinski said. 'And it's one hell of a lot lighter.'

'Let's work on that possibility.' Peterson held up his hand to stop further argument. He had seen from the slowness of the final assault that the team had been close to exhaustion and he could see it more clearly now in their eyes, the gauntness of their faces, the way they leaned on the ski sticks. For a moment he considered withholding his news until they had showered and changed, then he decided against. If they couldn't take the worst their own side could deliver, what the hell good were they going to be against the enemy?

'Now,' he said crisply. 'Most of you know Major Clifford already. But you may wonder why he's with me right now. The reason is that we have some news and it is not entirely good.'

He saw the questioning in Smith's expression, the apprehension of an ambitious man. The others were reacting more like grunts usually did. They showed a weary resignation to yet another change of plan, a 'word change' as they called it.

'This operation is being joined by the British Royal Marines commandos. That's the good news. They want to contribute a beach party and as many experienced mountain leaders as we care to ask for.' He glanced round the semicircle of men. 'I'm asking for one only. Not that I have anything but respect for the commandos. I just want to keep this whole thing small and beautiful. Now the bad news. For political reasons it's unlikely we'll mount this operation without full Norwegian participation.' He hesitated. There was no need for the men to know that Norway was already helping by pre-positioning equipment. The fewer people who knew that, the better. 'So there's no way you'll be going in yet and there'll be twenty-four daylight hours in Svalbard when you do.'

'Here we go again.' Burckhardt muttered. 'Hang around while those pogues in the rear shoot the shit.'

'Be clear, Burckhardt,' Peterson snapped, 'and all of you. I don't like this either. Though we do have to wait until the Soviets build the goddamn radar, after all!'

They all laughed and the tension subsided.

'Right. Now one result of making this more of a NATO operation is that it's upgraded. The British have assigned a landing specialist who happens to be a major.' He looked straight at Smith and spoke sympathetically. 'No criticism of you, Howard, none whatsoever. But we now have to put a major in command. We can't be outranked in our own operation.' He surveyed the team again. 'I say this in front of you all so there can be no misunderstanding. Major Clifford will assume command with immediate effect and have Captain Smith as second in command. Lieutenant Andrews, I'm afraid that puts you out.'

'Yes, sir. I understand, sir.' Andrews took it straight on the chin.

Peterson watched Smith closely. Despite the 'Duty, Honour, Country' credo inculcated during four years at West Point, despite the discipline, Smith was having trouble controlling his disappointment. He had gone to West Point on a football scholarship and never lost the feeling of being a star on the field. Peterson guessed that was why he had then opted for airborne and ranger training and finally volunteered for the Delta Force. He was a hero waiting for a battle and this raid was his big chance. Now he was only the number two, not the leader.

The seconds passed. Smith nodded but said nothing.

'Let's saddle up, then,' Peterson ordered. 'Burckhardt, call in the chopper, will you? We can get back to base.'

As the men broke away, Smith came across to him. 'I guess it's the way the cookie crumbles, sir,' he said reluctantly. 'As for the missile, I'm damn sure TOW's the answer.'

'Be certain, Captain, we'll evaluate every possible approach.' The way things are going, Peterson thought to himself, we're going to have plenty of time. We'll merely run out of snow to train in.

General Anderson's phone rang while he was dictating a draft to Sandra. The line was a secure link to the Supreme Allied Commander at SHAPE near Mons.

'Just hold it a moment,' he said to Sandra and took the call, listening attentively for a couple of minutes. 'They just have to agree up there. It's already announced the Marine Corps are involved in an unscheduled exercise. We have to send the ACE Mobile Force.'

Even though his words were being electronically scrambled, the General used as many circumlocutions as possible, and avoided naming the country. But the Allied Command Europe's Mobile Force existed to beef up NATO's presence on either the northern flank in Norway or the southern in Greece and Turkey, and he was determined it should be used now.

'We have to show the flag. Yes. I agree the Council prefers low-key measures. That's correct. We have to make a statement about practising deployment procedures. Sure, that's nonsense. They should be prepared to fight. If we don't send the Mobile Force, how else do we demonstrate support for a member country under threat?' Anderson spoke for a few minutes more, then put the phone down and turned back to his secretary. 'Now, Sandra, where was I?'

That evening Sandra was in a fret to get away on time. She was cooking dinner for Erich and worried in case she had to work late.

'Boyfriend trouble?' asked Julia, a new gold bangle

flashing on her wrist. She always sensed when Sandra had a special date. There was an underlying agitation in her manner which grew as the afternoon wore on. 'They're none of them worth it, dear. Take my advice. Kick them in the teeth a few times. Keep them in order.' She gave the Sergeant a heavily mascara'd wink. 'Krauts especially. They like to be disciplined.

'Don't take any notice of me, darling, I'm just jealous, that's all. That stinking man of mine's gone back to his wife.'

'Oh, I am sorry.'

'They come, they go.' Julia wrinkled her nose impishly. 'That's the lovely thing about NATO. New men all the time. Why d'you suppose I've stuck it all these years?' She gave a drawling emphasis to the phrase. 'All' was one of her words. 'Anyway, don't you worry. If there is a crisis at five, I'll stay late. I owe you a favour, don't I?'

It was a curious thing about Julia, Sandra thought as she drove back into Brussels after work. You'd imagine she was so man-mad that she'd never remember anything. In practice, far from being scatter-brained, Julia had an astute ear for what was going on, and Brigadier Curtis was not the first Chief of Staff to rely heavily both on that and on her knowledge of how the organization functioned. Sandra hated the idea of featuring in the perpetual NATO gossip. The one catch about working here, for all the high pay and good conditions, was that the need for secrecy made everything feel inbred and promoted intrigue.

That evening Erich arrived punctually, though in a grumpy and perverse mood. He downed his first whisky with unusual rapidity. Nor, to her disappointment, did he notice that she had rinsed her hair several shades lighter. Half afraid the experiment would be a failure, she was going blonde slowly. But he was so clearly upset that she concentrated on consoling him.

'What's the matter, darling?' she asked sympathetically, as she made some salad dressing. She was never self-confident enough to finish the preparation of a meal completely before a guest arrived, and liked an excuse to bob in and out of the kitchen in case something burned.

'I heard today,' he said, '*Der Spiegel* will definitely be looking for a new man in the autumn. That gives me five months to make my name. God in heaven, I need some scoops soon.' He clasped his head in his hands as though vexed beyond bearing. There was no need to mention NATO. She knew well enough what story he wanted, though he used the banal word 'scoop' to remind her. If she didn't come up with information soon, he was going to have to start exerting pressure, serious pressure.

'There's really very little happening,' she said, and turned the conversation to another topic. But her mind kept reverting to General Anderson's phone call. She debated it while she lifted the steaks from the grill. There was to be another exercise in Norway and Anderson had specifically mentioned an official statement. Surely it could do no harm to tip Erich off to expect something. He was so disconsolate, poor man.

When they reached the coffee, and a couple of glasses of good Beaujolais had helped her relax, she made her decision. Why not? It couldn't possibly do any harm.

'Actually, darling, I did overhear one tiny thing which might help you. There's going to be a statement soon about an exercise in Norway.'

'There is?' He was immediately alert. 'But the Polar Express is already over. They have issued photographs.'

'Well, I think they're following it up with another.'

'And will that impress the Russians?' He allowed himself a flash of scorn. 'Every picture from Polar Express shows soldiers eating or drinking or sitting in the sun! Exercise Polar Holiday, we call it.'

'I don't know about that.' She didn't like his sarcasm, even though there was truth in it. 'Anyway, the point is you should have a story in a few days.'

'That is good,' he agreed hastily. 'Thank you, my love, thank you for telling me.' He clinked his glass against hers and drank a silent toast. Her eyes were full of pleasure. Yes, he thought, this is a moment to drink to. A historic occasion: the first time she had volunteered military information. 'About our little cross-country flight, sweetheart. Have you ever been to Cannes?'

'Wouldn't that be terribly expensive?'

'Deauville is closer.'

'Let's go there then. It's still France, after all. I do so love France, especially in the spring.'

'Deauville it shall be.' He squeezed her hand under the table. 'We will go in May, when the weather is warming up.' And when the confrontation with NATO would be hotter too, he thought, smiling at the secret joke. She held his hand and smiled adoringly back.

On a hill above Bardufoss airfield the Norwegian anti-aircraft gunners stood by. Their Bofors guns were half concealed in snow pits. A few yards away rose the pallid dome of the radar which fed information to the guns, tracking approaching aircraft. Inside the warmth of the radar cabin, the crew monitored a steady stream of approaching aircraft. The ACE Mobile Force exercise had begun and C-141s, huge long-winged jets, were bringing troops from the United States and Canada; stubby Hercules turboprops had already offloaded a British battalion from Tidworth in Wiltshire; the force commander had flown from his headquarters in south Germany yesterday; more troops were arriving all the time.

Down on the airfield administrative officers were calmly bringing order out of potential chaos; documenting troops in a large hangar after they had disembarked, sending them to temporary holding areas cut for this eventuality from the woods within the perimeter. Tents were being pitched, meals cooked in huge aluminium pots, defences set up by the black-bereted crews of a Leopard tank squadron permanently based here.

From Bardufoss the international force would move to exercise areas in Troms county, but all the time their commanders would be studying maps of Finnmark, the extreme north-eastern part of the country where the frontier with the Soviet Union stretched from the Barents Sea inland for 191 kilometres to the Finnish boundary. Not that the Russians would restrict themselves to movement across the common frontier if they did invade. Finnish neutrality was not expected to withstand a Soviet demand for passage along roads which led to Norway.

The Mobile Force were not the only troops preparing. Norwegian conscripts were being rushed north in chartered airliners to strengthen the North Norway Command. The Home Guard was briefed: every bridge in the north had demolitions preplanned for it, every important road junction its ambush, every power station its defence. Although the Prime Minister, in his broadcasts, stressed that there was no immediate danger, all Norwegians knew that a crisis was approaching. As the Allied snowmobiles and trucks rumbled north from Bardufoss, the villagers lined the roads, waving flags and cheering.

In Moscow, *Pravda* published a long editorial denouncing the imperialist NATO exercise as endangering Scandinavian security and threatening world peace. 'It serves to confirm the necessity for the USSR to participate in the defence of Svalbard's neutrality,' the paper declared.

'Makarov is obsessed with the mountain. I am sure it will go up there.' Folvik was repeating himself, struggling with the implications of what Mydland had just asked. To send clandestine radio messages back to Norway, to be a spy, possibly to be stood against a wall and shot, to be a hero. The images raced through his mind, his mother weeping by his grave, the priest standing in the snow, photographs in the papers, he saw it all as though in a television thriller. Then he shook his head, unbelieving. The events of the last two months were unreal. First the Russians seizing the airfield in which he took such pride, the humiliations of sharing duty with them, of taking orders from the arrogant Makarov, and now this quiet-spoken fellow Norwegian inviting him for a chat and putting the most dangerous request ever made of him. 'Surely, if they build anything military, they will replace us with Russians?'

'In the end,' Mydland conceded. The General had briefed him well on this aspect. 'But how long is a piece of string? No one knows what will happen. We must get information while we can.'

Folvik sat silent, sipping his coffee. For once the mild stimulant gave him no sense of wellbeing.

'I repeat,' said Mydland softly. 'No one will think the worse of you for refusing.'

'All right.' The words took a long time coming out. 'I'll help. But where do we keep the radio?'

'Annie will hide it.'

'I must also know how it works.' Now there was a stubbornness in Folvik's tone. 'Something might happen to her.'

'It would be safer simply to leave messages for her in a prearranged place.'

Folvik shook his head. 'No,' he said. 'Everyone is in everyone else's pocket here. Better for me to continue meeting her casually from time to time, as I do now.'

Mydland did not press the point, the duration of the arrangement might be very limited anyhow. He showed Folvik the functioning of the radio and how to feed in the encoded words before making the actual transmission. He stayed behind when Folvik left, not venturing out for several hours. The remark about Longyearbyen's gossip had struck home.

The next day Folvik heard a rumour which made him tremble all over and tested his resolve to the limit. The story was that Russian police had arrested Paul Mydland and another man and accused them of being spies.

5

'As Norwegian citizens, employed by the Store Norske, they have every right to be here.' Sysselmann Prebensen spoke quietly, with all the conviction he could muster. He was facing Stolypin across the long, black-stained table in the conference room. Outside in the passage Mydland and Haakon were being guarded by four Russian border police.

'They are spies,' remarked Stolypin flatly, the interpreter echoing the almost casual authority of his words. 'They did not come direct from Svea, they have not been making a survey. You are lying, all of you.'

'I object to that accusation.'

Stolypin did not reply immediately. He reached out and picked up the miniature Norwegian flag which stood alongside the red Soviet one in the centre of the table, reflecting on reactions he had already planned. His impassive self-assurance was the mark of a man who had been at the top for many years. He gazed at the flag, with its blue and white cross on a red ground, then with one finger rolled it up around its tiny pole. 'Is that what you want?' he asked.

The Sysselmann looked steadfastly back at the Russian, controlling his temper. His orders from Oslo were to maintain the status quo at all costs. He was the only official in Svalbard aware of Mydland's double role. Now it was politically vital to get the two men out.

'There is a plane due from Tromso tomorrow,' he said. 'They can go back with it.'

Stolypin played with the toylike flag, his stubby fingers rolling and unrolling it in a way that angered Prebensen. 'In our country we shoot spies.'

'In ours – ' the Sysselmann dwelled on the words ' – a man is innocent until proved guilty.'

The reprimand was undisguised. The Russian's lined face hardened. 'This is a deliberate provocation by your government,' he said.

Prebensen looked squarely across the table, his long jaw set. His remark about justice had evidently struck home or Stolypin would not have shifted his ground. Weak as his own position was, he still had cards to play, the chief one being the Soviet Union's presumed desire to maintain this joint government. However much the General was trusted, could he risk a premature breakdown of this arrangement? Prebensen guessed he could not.

'If you do not trust my government,' he said, 'then I have no option but to withdraw myself. You can haul down the Norwegian flag throughout Svalbard.'

Stolypin grunted but appeared unflustered. 'I do not mean these things personally,' he said. 'For myself I have faith in your integrity.'

Prebensen inclined his head in an almost mocking acknowledgment. 'I'm glad of that,' he said sardonically. 'However, I must insist on my right to deal with the activities of Norwegian citizens as I think fit. Just as you deal with Soviet subjects?' He left the rest unspoken. It was obvious: either Mydland and Haakon leave or I do.

'And where will these two ... "citizens" ... stay tonight?'

'At my house.'

'They should be under arrest.'

'Charged with what? Their papers are in order.' Prebensen knew he had won, though it was hardly a threat he could expect to repeat. Accordingly he tempered victory with conciliation. 'I give my personal guarantee that they will be at the airport tomorrow.'

'Very well.' Stolypin replaced the little flag on its mount. 'I accept your assurance. Now, there are other matters to discuss. The airport, for example. We need another hangar for Aeroflot. . . .'

He was going to exact a price for releasing those men, though he still wished he knew how they had spent the thirty hours between leaving Svea and reaching Longyearbyen. 'Other facilities also need improving. I shall want your full collaboration.' Yes, the price would be high.

A week later a section of Russian border police were reconnoitring around the trappers' hut by the Van Mijenfjord when they found traces of the *skijeger*'s stay, the few indicators even the most watchful may fail to remove: a cigarette butt, a scrap of packing from the rations, most damning of all, a tiny patch of oil in the snow. The oil had been covered by snow, then revealed again when a gale blew off the fresh fall. When the lieutenant leading the patrol saw it, he began examining the whole surface. Disregarding the biting cold, he scooped away snow with his mittened hands until he came to the hard packed place where the Sea King had rested. Making his men excavate as patiently as if they were archaeologists, he traced out the pattern of the marks and knew for certain a helicopter had landed. What he had not been able to tell was how many men had been there, or where they had gone. The ski trails had totally disappeared.

Makarov was busy executing a sketch for Stolypin's benefit when a rap on the door disturbed him.

'Enter,' he called out.

The young Russian from the control staff, Paputin, came in, his face alight with excitement. 'Comrade Colonel,' he blurted out. 'One of the patrols has found signs of military activity on the Van Mijenfjord. At a trappers' hut.'

Makarov rose to his feet and consulted the map on the wall. 'They have a reference?' he demanded.

Paputin handed him a slip of paper, then stood back deferentially.

'Tell them to leave four men there and the rest to return.'

Dismissing the controller, Makarov remained on his feet studying the contours on the map. The General was going to want a reason for men having landed there. It seemed inescapable that it involved the two men they had released.

Frederik Folvik had come on duty at midday, pursuing the normal routine of taking over: noting weather conditions and the atmospheric pressure reading on the chromium-plated aneroid barometer made by Negretti and Zambra of Croydon, England; checking the few anticipated arrivals and departures.

'There are also two Soviet helicopters out,' remarked his Norwegian colleague. 'No estimated time of arrival back.'

'You need not concern yourself with them.' Paputin had been listening. 'Your concern is with commercial traffic only.'

'But we need to be advised of all movements,' Folvik protested.

'These do not concern you!' The Russian was curt to the verge of anger.

Folvik caught a glance from his compatriot and did not pursue the point. The glance told him clearly that something was afoot today. When the handover was complete he boiled the kettle and made himself coffee, deliberately not offering any to Paputin. Cooperation can work two ways, he thought to himself. A moment later it occurred to him that he ought to find out what was going on, so he gave him a cup. But the gesture failed to pay off. Only much later, when an MI-6 helicopter clattered down out of the sky and disgorged twenty armed police, did he gain any clue to what was happening. Paputin looked down from the long windows and one of the police, an officer with a sub-machine-gun slung over his white camouflage suit, raised an arm in acknowledgment and stuck a mittened thumb high in the air.

In the evening, pondering these events, Folvik realized the full significance of what he had seen. Normally the Soviet police wore long greatcoats, with green tabs on the lapels and peaked uniform hats. The men who had jumped in quick succession from the helicopters had all been in snow fighting kit, with white hooded oversuits. They could only have been on patrol up in the mountains. If so, what had they found?

He hurried out from his own apartment to the block where Annie lived. An hour later they sent his second 'burst' message on the clandestine radio. The first had concerned the assembly of equipment in the hangar. By mutual agreement, they only used the set when something of great importance was taking place.

The next morning, General Stolypin, coldly furious at having been outwitted, sent an urgent request to Moscow. He wanted to be assigned a detachment of the *Spetsnaz-nacheniya*, the Special Designation units who were the élite of the KGB's troops. It was the Spetsnaz, as they were colloquially known, who had seized key points and arrested Premier Dubcek in Prague in 1968 and who, a decade later, secured the airfields in Afghanistan before the Soviet invasion. Good as the Border Troops were, the Spetsnaz were better and Stolypin wanted the best. As he explained in his signal, the task of safeguarding Svalbard might appear simple on paper but in practice it was extremely demanding.

The reply promised immediate dispatch of Special Designation troops, who would be flown in as 'mining workers', wearing civilian clothes. They were to be quartered at the Russian settlement of Barentsburg, well away from Norwegian prying. Nor was Stolypin surprised at being ordered to bring forward erection of the surveillance radar on the Plata Berget to the earliest possible date. He had guessed that the Politburo would react to the provocation of the new NATO exercises with positive action and he wondered what other steps would be taken.

The watchtower was constructed of dark, creosoted wood. It stood on a hill overlooking the frozen Pasvik river, this morning invisible beneath a shroud of fog. Inside the tower two Norwegian border guards, both young conscripts near the end of their service, shivered and watched the pale orange sun creep up out of the haze until its light suffused the cotton-wool layer of mist. A series of dark fingers became visible in the distance, smoke drifting from them: the top few feet of the factory chimneys of Nikel, the Russian mining town across the valley.

As the two Norwegians gazed across the whiteness a straying reindeer took shape near the foot of the tower. It began pawing the snow to uncover the moss hidden below, then suddenly stopped and raised its head, antlers quivering. The guards heard the noise from the hidden valley too.

'Something's happening out there,' said one.

'Probably a patrol.' The Russian border police often checked their 119-mile frontier with Norway. Not that there were many intrusions, except by animals. But this was the Kola Peninsula, the place where the Soviet Union had the largest concentration of military power in the world and also the only section of frontier apart from Turkish Armenia where the USSR abutted on a NATO country.

The reindeer shifted, its ears twitching, disturbed further by the guards' subdued voices. Then it moved off down the slope, its brown outline soon swallowed by the mist.

An hour passed. The sun rose slightly higher in a sky streaked with cloud and the fog began to disperse, revealing parts of the frozen river and the leafless birch trees along its banks. The far side came into view.

'What in hell!' The corporal raised his binoculars to his eyes, astonished. Facing them, away across the ice, were the squat bludgeon shapes of three tanks. The thick-barrelled guns in their turrets seemed strangely foreshortened. Suddenly he realized why. They were aimed directly at his tower. For a moment he was seriously frightened, then he controlled himself and picked up the telephone.

At the Border Commissioner's headquarters in Kirkenes, the Norwegian town a few miles back from the frontier, the messages came in fast as the fog cleared. From the observation posts of all seven border stations the reports were similar. The Soviet Army was ranged along the frontier in force.

By midday the Commissioner, a colonel, had reported to Oslo that he was confronted by at least a motor rifle division. Like the Sysselmann in Svalbard, he was responsible to the Ministry of Justice. Unlike the Sysselmann, he had a formalized procedure for discussing frontier questions with the Russians, a procedure which cut across the 3000-mile-long diplomatic circuit via Oslo, Moscow and Murmansk. He used a red telephone with no dial which communicated direct with his Russian equivalent, a KGB colonel. The conversations on the red telephone were automatically recorded by a Stenorette machine attached to it. But when he had finished, all the tape held was a gruff interpreter's reply that a meeting would not be convenient.

The colonel, aided by a junior officer, resumed indicating the positions of Soviet troops on a large-scale map. He knew that the Russians had last massed troops on the frontier in June 1968. In August they had invaded Czechoslovakia. Since then, official Norwegian thinking had veered to saying there had been no connection between the two events. The colonel didn't believe that, and he had no doubt whatever that the tank detachments his assistant was now marking with red symbols on the map were a direct response to the arrival in Norway yesterday of more American Marines. In his view it was a clear warning to NATO to keep out of the Svalbard dispute.

*

In West Germany, at the sprawling British Army of the Rhine headquarters near Monchen-Gladbach, the senior Royal Signals officer studied reports from his electronic counter-measures experts. The volume of military radio traffic in East Germany, monitored from many different listening posts, had increased sharply during the night. A few hours later the intelligence staff were collating translations of the radio messages with satellite photographs. The Soviet divisions in East Germany were on the move and making no secret of the fact.

'You know,' remarked the Commander-in-Chief with very English humour, 'I believe they're trying to tell us something for a change.'

'I agree,' said his political adviser, a balding civilian detached from the diplomatic service. 'And in my opinion the message is that if we send one more soldier to Norway, they'll cross into West Germany.'

'And?'

'We should send that one soldier immediately, because this is bluff of the crudest kind. But we won't, will we, General? Because NATO won't have the nerve.'

In the huge Dutch port of Rotterdam, proud of being the gateway to Europe, a correspondent of the Soviet news agency, TASS, was sitting in a bar with a friend. The two men were drinking beer chased by gulps of fiery genever, the local schnapps distilled from rye and malt. Beer and genever are a traditional, and lethal, combination. The Russian, Yuri Vassilevich Gamov, though barely half the age of his companion, a union leader, was determined to prove his capacity and the mixture was going to his head.

'If Moscow decides it,' he boasted, '50,000 Dutchmen will take to the streets.'

'Without us the dockers won't.' The union leader was a burly man, his beer-fed paunch bulging above his belt. He sat with both elbows squarely on the table. 'Who d'you think you can rely on to get the workers out, the priests?' He looked balefully at the young Russian. There had been a lot of scandal recently about church activists being involved with the KGB in organizing anti-NATO demonstrations.

'Those who believe in disarmament will march: students, schoolteachers, conscientious objectors.' Gamov was being forced on to the defensive, much against his will. He was supposed to be giving orders at this meeting, not answering questions.

'Dutch Marines training in Norway is a different thing from the bomb.'

'They're both a threat to peace. I tell you the whole of Holland should protest.'

'A few intellectuals might do.' The contempt rang in the docker's voice. 'But you won't get the workers without us.' He leaned forward heavily, lowering his gruff voice further. 'Not that they wouldn't like a little entertainment on a Sunday. So long as they're not losing overtime.'

The Russian flushed, annoyed at the conversation veering towards money so rapidly. 'There would be a contribution to union funds,' he said and gulped down more genever, trying to give finality to the statement.

'It had better be more generous than last time.' The Dutchman stared back hard and insolently.

For a moment Gamov almost lost his temper, but even as he choked back a retort he felt a sudden wave of nausea in his stomach. The room went momentarily out of focus. He had to take breath to recover his composure, realizing with a sick feeling that he had drunk too much, and that the man opposite knew it. 'The contribution will be substantial,' he conceded. 'These demonstrations are important.'

'We could drink to that then.' The docker smiled slyly, wiped the back of his hand across his lips, and raised it to summon a waiter, the cloth of his coat straining across his broad shoulders. He moved with the ponderousness of an old bear. When he caught the waiter's attention, he glanced at Gamov. 'How do you like our genever? Sorts out the men from the boys, eh?'

Gamov nodded reluctantly. He had just remembered being warned not to try drinking a Dutchman under the table. 'Even a Russian may fail at that,' his friend had said. Failure did not feature in Gamov's brief from the KGB on the organization of protest in Holland.

A crowd gathered in London's Trafalgar Square at lunchtime, where the grimed stone pillar of Nelson's Column soared above the men and women in jeans and sweaters, many wearing anoraks against a chilly wind. They held aloft banners which read 'War Out, Peace In', 'Quit NATO Now' and 'Marines Go Home'. The inverted Y in a circle of the Campaign for Nuclear Disarmament was everywhere, on the banners, on demonstrators' lapels, on leaflets being handed round by young, leather-jacketed organizers. Small groups of police stood by around the edge of the square, waiting for the march to begin. To march with banners required official permission and none had been requested.

The last speaker climbed down from his improvised platform by one of the great bronze lions which guarded Nelson's Column and the crowd began to shift, marshalled by the cheerleaders. A rhythmical chant started. 'NATO out, out, out. Peace in, in, in.' The banners and placards waved above the crowd as it began to surge down Whitehall, forcing the traffic to stop, heading for the Prime Minister's residence in Downing Street.

From a control van parked near the Admiralty Arch, a police superintendent gave orders. The blue-helmeted constables moved, cutting into the developing line of marchers. Immediately there were scuffles, fisticuffs, kicks. A girl in jeans, who would never normally have used a hatpin in her life, drew one like a dagger from a folk-weave bag and jabbed at a police horse's flank. The animal whinnied and reared, its hoofs flying, coming down on two men, as the rider whirled his polished wooden truncheon. The crowd roared angrily. More police ran out of a sidestreet, this time with riot shields and batons. The crowd became a mob, banners were trampled, women screamed. It took three hundred police over an hour to disperse the demonstrators and there were seventeen arrests.

'Spontaneous,' commented one constable, wiping the blood from his face in the washroom of Cannon Row police station afterwards. 'That was about as fucking spontaneous as Christmas.'

The same afternoon, in the ornate Gothic chamber of the House of Commons, left-wing Members of Parliament took advantage of Question Time to attack police brutality and demand an emergency debate on NATO's warmongering confrontation with the Warsaw Pact.

The snowstorm blew up with a rapidity which caught Makarov unawares. He was up on the Plata Berget with young Yakushkin, the radar engineer, supervising the erection of the 80-foot-high mast for the surveillance radar. He was a good man, Pyotr Ivanovich Yakushkin, sensible. Makarov had warmed to him after the General's inspection of the site in April, when he had told Stolypin respectfully yet firmly that to attempt the final assembly of the radar in the appalling conditions of the mountaintop was impractical. Despite the urgency, the preparations had taken a month.

The aerial was the kind used in the defences of Murmansk, a monstrous affair looking like part of a suspension bridge. From each side of the mast the cables stretched out to support a 120-foot-wide horizontal section which on a bridge would be the roadway, except that below it was a mirror image of the suspension cables above, while between these cables and the 'roadway' were numerous vertical pieces of aluminium: the all-important dipoles from which the radar waves were transmitted. Western radar experts jokingly called it the 'Forth Bridge type'.

However, this radar was well proven in conditions of extreme cold. Once erected on its base, which Makarov's workmen had anchored in the rock with steel stanchions, and rotating every twenty seconds, it would track anything in the sky up to 400 miles away. A cable snaking down the mountainside would take this information to a control centre on Longyearbyen airfield, where computers would automatically sort out its significance. From the control centre microwave transmissions routed via a satellite stationary above the equator would beam the signals to Archangel. Both the radar waves and the microwaves travelled one nautical mile in a fraction over twelve and a half millionths of a second: 12:63 to be precise. So this seemingly grotesque aluminium lattice could transmit identification of an American aircraft 400 miles out over Greenland to the defenders of Murmansk in four to five seconds. During that time a B-52 with a cruise missile on board would have come all of one mile closer to Russia. Even though there were already four-engined TU-126 aircraft, burdened with airborne early warning radar, on permanent patrol over the North Atlantic, the surveillance radar on Plata Berget gave a great advantage to the Soviet defences. Taken in conjunction with the 2000-mile-range radars soon to be erected at the mouth of the Isfjord, the advantage was as vast as anything the Americans feared.

For this reason Stolypin had eventually given way to Yakushkin's insistence that the only way to place the radar up on Plata Berget was by bringing one of the Red Air Force's giant crane helicopters and lifting the whole thing up in one piece from the airfield. The General, following his orders from the Politburo, had wanted to present the world with a *fait accompli*: to have the radar working up there before other signatories of the Treaty had time to protest. He feared that since it was too large to be assembled secretly inside the hangar, Norwegian spies would inevitably report its existence back to Oslo and surprise would be lost. However, Yakushkin's logic was irrefutable. A hundred calamities were possible if they tried to assemble it up on the mountain. Makarov was privately delighted at this victory of common sense over politics and from that moment had regarded the mild-faced, mousy-haired Yakushkin as an ally.

The decision to erect the radar today had been taken at seven this morning, when the sky was a clear pale blue and the forecast augured well. So now, at midday, Makarov and Yakushkin were standing in the snow on the Plata Berget, near the hut that had been put up to shelter the construction workers, watching as the huge helicopter manoeuvred towards the site, the monstrous aerial swaying beneath it. They were swathed in sheepskin coats, their legs protected by Army high boots made of rubber lined with felt, and Makarov held a megaphone to direct the workmen who would guide the bottom of the mast into its socket when the helicopter was hovering overhead. He could also talk to the pilot on a ground-to-air radio. He was concentrating so much on the task that he was unaware of the snowstorm behind him until a sudden, tearing gust of wind cut through his clothing and rocked the helicopter. He swung round and saw a white curtain of snow obliterating the view of the fjord, though still a couple of miles off.

'Hey, Pyotr Ivanovich,' he shouted. 'Look left.'

Yakushkin turned and Makarov saw the dismay on his pale face.

'How long do we need to secure it?'

'Half an hour.' Yakushkin hesitated, eyeing the helicopter slowly approaching, the radar swaying perilously in another gust. 'Send it back,' he yelled. 'This is not safe.'

Any fool could tell that, thought Makarov, but could he send the helicopter down again? For a long moment he studied the nearing storm. God in heaven, this cursed place was unpredictable. The weather changed in minutes. With a new apprehension he realized that the whirling snowstorm was if anything moving along the fjord. It would envelop the airfield as soon as, if not before, it reached them here.

The pilot could not fly back, not into a total white-out. He raised the radio microphone to his mouth.

'Can you expedite?' he asked the pilot calmly. 'Do you see the storm?'

'I will try.' As the pilot replied Makarov saw the helicopter surge forward. He turned the megaphone towards the workmen of the construction detachment, stumpy figures in quilted khaki suits. 'Stand by to secure the mast!' They ran towards the huge dangling aerial, stumbling in the snow, trying to seize the guide ropes attached near the base.

The gusts of wind had become almost continuous now, the white wall of the blizzard was nearer and the first flakes stung Makarov's cheeks like spikes of ice. He could no longer distinguish the far edge of the plateau. The helicopter was a hundred feet up, rocking violently, its angular shape black against the sky, the crewman leaning out perilously as he directed the pilot. But the workmen had their hands on the ropes and the mast was only feet from its socket.

'Steady,' he warned over the radio.

The wind rose further, becoming a gale, and Makarov realized he had misjudged his distances. In seconds the landscape was completely blotted out, and the helicopter reduced to a heavy shadow heaving in the whirling, driving snow. He heard the crewman, his voice overlaid with fear, telling the pilot to climb. Idiot, thought Makarov. It was any helicopter crewman's instinctive reaction to danger, to climb away. But with this load to lose sight of the ground would be disastrous.

'No,' he shouted into the mike. 'Lower, bring it down.'

'Cut loose!' came the pilot's voice. 'I can't hold her.'

'No. Lower it!' Instinctively Makarov stumbled forward, clasping the microphone, straining to see the gaunt shape of the helicopter. The bottom of the mast was dragging badly, ploughing a twisting furrow in the snow, while the workmen frantically tried to hold it steady. He shouted at them to let go completely, but his command was lost in the roar. One had the sense to do so. For a few seconds he believed the mast could be saved and then, through stinging eyes, he saw the helicopter shuddering. The combined weight of the mast, those fools pulling on the lines and the force of the gale were too much. The tail dipped savagely, there was a terrible, grinding, jangling impact and the tail rotor sliced into part of the aerial near the mast. A scream came over the radio as the huge machine cartwheeled and the crewman tumbled from the open door to hang on his safety strap below. The image froze in Makarov's mind. The machine grotesquely keeling over, with the man hanging like a doll on a string. Then the main rotor cut into the mast itself, and the helicopter tumbled upside down into the snow, metal tearing and rending, until it hit the ground and exploded.

Makarov threw himself flat, fragments whining past him like shrapnel, then the noise subsided into a tremendous hissing as burning fuel fought against snow. He stood up choking in the acrid smoke and shouted hoarsely to the nearest workman to fetch fire extinguishers from the hut, though he knew nothing could save the helicopter crew. The wreck was an inferno.

'So, how do you like the place, sweetheart? You must admit it could be worse.'

Nancy Peterson was showing off the small apartment she had rented in South Kensington. She had carried out her threat to follow her husband to London if he was gone more than a week or two and he had been operating out of the US Navy's European headquarters in Grosvenor Square for nearly a month. The raid was ready to be launched. The Delta Force team were getting to know their Royal Marine commando colleagues in a remote Scottish training area. But although reconnaissance revealed foundations being laid on the Plata, no radar masts had appeared and the Norwegians were hoping against hope for a diplomatic solution to the crisis. So they all waited and the team's morale began to suffer.

'It's better than a hotel, surely?' she insisted. The block was a refurbished 1930s building near the underground station and the apartment itself had a living room, a kitchen the size of a galley in a yacht and a double bedroom scarcely larger.

'How in the hell did they get furniture in here?' Peterson asked laughing. He noticed the pink roses she had arranged in a vase on the table and the bottle of Kentucky bourbon with glasses and ice on a tray. 'You're the darnedest woman.' He kissed her, then swung her round in his arms happily.

'Careful!'

'Oh Jesus, I keep forgetting.' He let her go and shook his head, his face crinkled in amused tolerance. 'Well, you said you'd come across and you damn well did, baby or no baby.'

'You're pleased?' The agent who had leased her the place swore she was lucky to find anything in May, with the London season reaching full swing in a week or so. 'It's close to the subway and you go straight through to Mayfair on this line.'

'It's great.' He kissed her again. He was both delighted and troubled that she had carried out her threat and come. The snag was, he might not be here long. Not after today's news. He had been called down from Scotland this morning. At the Navy headquarters they had satellite photographs which were making everyone think hard.

'You say it's great, but you look as if the roof just fell in. Tom!' She gazed at him beseechingly. 'They're not sending you some other place, are they?'

'Not that I know of.' Hell, that was true. He had no idea what would happen next. The satellite photographs had shown the burned wreckage of the radar and the helicopter up on the Plata Berget. As one of his colleagues had commented, 'Seems like they've done a pretty good job on it themselves, Tom. Hardly necessary for us to step in and help them yet awhile.' However, the Russians were bound to start again, so the raid was not abandoned, merely postponed, and he was to keep the team in training.

'I may be up in Scotland pretty often,' he said. 'We have a programme there with the British Marines.' It was as much as he could tell her. 'Hey, sweetheart, what are we talking about work for? We should be celebrating. Let's break open that bourbon for a start. You know a nice place to eat around here?'

'I saw an Italian restaurant down the road.'

'Let's give it a whirl.'

'All in good time, Tom.' She bit her lip, annoyed with herself at being abrupt, but he was as restless as a cat and all too obviously had things on his mind. 'Let's just sit quiet for a moment and catch up with ourselves, for heaven's sake. I haven't seen you for a month, remember?'

'Sorry, honey.' He poured two generous measures of the whisky and sat down in an armchair, telling himself to relax. A couple of simple questions about how things were back in Virginia Beach started Nancy recounting the way she'd obtained leave from teaching this term. He half listened, absorbing some of the news, though simmering in his mind all the time was where things were going with the raiding party. It wasn't just the problems of morale which postponements raised. The more he watched that second in command, Howard Smith, the more doubtful he was of his ability to temper ambition with obedience. Of his courage there was no doubt, but when the crunch came he made his own decisions.

Yesterday's final exercise with the British commandos, conducted in a remote area of Scotland where the local people were accustomed to live firing, had begun with the team slipping ashore from rubber boats in a loch and culminated in a variation on the previous attack plans. Craig Clifford had decided to utilize the lighter Dragon missile to create a diversion and sent Smith and Trevinski off through the dank early morning mist to one flank to attempt this. The idea was to give the enemy, played by commandos dug in at a safe distance from the target, the impression that the attackers had surrounded them. But when the mist suddenly began to clear and Smith found himself only 500 metres from the 'radar' and in an exposed position he had opened fire without obtaining a confirmatory radio order from Clifford. The radar mock-up had been set on fire before Clifford himself was ready to attack. The defending commandos had leaped from their cover in hot pursuit. Since they were not firing live it was arguable whether the team had been annihilated or not.

'Well, Colonel, we zapped that radar pretty good,' Smith had argued at the debrief afterwards, 'and if I'd hung around waiting they'd for sure have zapped us. Didn't have too much choice when that fog cleared.'

It was a finely balanced argument when you trained men to take the initiative. On the other hand Smith had been ordered to wait for Clifford's command.

'Let's lay one thing on the line, Captain,' Peterson had said to him privately afterwards. 'I don't want an action replay of Custer's Last Stand. I want you taking out that installation and being lifted out after. Like I said before, there isn't going to be any big medal for this action.' Smith had flushed at the reprimand and remained stonily silent.

Reflecting on it now, twelve hours later and a world away in atmosphere, Peterson feared he had been too harsh.

In any case the whole damn project was full of new uncertainties. The satellite photographs of the burned-out radar on the Plata Berget had shaken everyone concerned. The intelligence experts had supposed the radar would be inside some kind of weatherproof geodetic dome. But the tangled metal lying there now had been an open-air structure, and one which could only be demolished by a phenomenally accurate shot with any missile. The conclusion was that they'd have to fight their way right to the radar, if the same kind was successfully erected, and place charges around its base. Unless there were very few defenders, such an attempt would be suicidal. Finally, compounding this uncertainty, there were signs of new construction near the existing Norwegian radio station on the Isfjord, an obvious site for long-range radar. If one was blown up, would the Soviets abandon the other? The psychology of the Kremlin's reactions was for generals and politicians to judge. But Peterson didn't want to see good men lost for nothing.

'You certainly have something on your mind, Tom.' Nancy cut into his thoughts sympathetically now, because she guessed something was badly wrong. 'Can't you tell me what it is? Maybe I could help.'

He screwed up his face, grinned and shook his head. 'I had a mother of a day, it'll sort itself out. Let's go eat soon, shall we?'

Later, in the lingering warmth of the London evening, they walked around the neighbourhood, exploring. It was a tenet of Peterson's that he oriented himself as soon as possible in a new location. The restaurant was down the Old Brompton Road, so they wandered along past antique galleries and boutiques, catching glimpses of elegant stuccoed squares and gardens. Yet overlaying the sophisticated surroundings of Kensington, there was an air of apprehension.

'You know something,' Nancy commented. 'There is an amazing amount of graffiti around.'

'So there is,' Tom agreed. Just about every blank stretch of wall was spray painted with nuclear disarmament symbols or pasted with crudely printed fly-posters demanding Britain get out of NATO.

Ahead of them a young man in a T-shirt and jeans stood on the pavement. He thrust out a leaflet, then eyed them with hostility. 'You American?' he demanded. He knew they were. Both the man's short hair and the cut of the woman's dress shouted it aloud.

'That's correct,' Tom replied easily.

'You should get the fucking hell out of Europe.'

'Just what do you mean by that?' demanded Nancy, her cheeks flushing with anger.

'Fucking warmongers . . .' The youth felt like spitting in

their faces. He saw the line of Peterson's jaw, the scar and the hard blue eyes, and thought better of it.

'Listen, son,' said Tom, his fists clenched at his sides. 'You want to talk to us, lay off the swearing, right? My wife doesn't like it. Nor do I.' Out of the corner of his eye he could see another man with a bundle of leaflets crossing the street. Cool it, he told himself.

'I don't want to talk to capitalist pigs,' shouted the youth, backing a pace away. 'You don't want peace. What right have you to be here anyway?'

'Let's go,' said Tom, as much to the boy as to Nancy. 'Otherwise I'm going to lose my temper with this little bastard.' He took her arm to guide her past, but the second man was in the way. He was broader-shouldered than the first and wore a black leather jacket decorated with chromium studs.

'You insulting my friend?' said the second man in a surly voice. He stood with legs slightly apart, blocking their path.

'I'll give you five seconds to get out of our way.' Peterson let go of Nancy's arm but his eyes never left the thug. 'Just stand back, honey. One . . . two . . . three . . .'

'Come on, Ted.' The leather-jacketed one's nerve broke.

They moved off down the street. A few yards away the younger one looked back and shouted, 'Pigs, fucking pigs!'

'I think I need a drink after that,' said Nancy, still holding on to him. 'I just never imagined England could be like this.'

'It ain't!' A tubby man in a colourful zipped-up pullover had come up unnoticed behind them. 'Take no notice of that sort,' he said. 'They're just the lunatic fringe.'

Peterson took in the man's creased and friendly features. He looked like a repairman on his day off.

'There's a good pub close,' the man said, 'if you feel like a quick one.' He led the way to a Victorian public house, all mahogany and cut-glass mirrors, with a smoke-stained ceiling, and insisted on buying them a drink. 'I'm Labour born and bred,' he remarked cheerfully. 'Been on the council sixteen years, not that we ever get control. Solid Tory this part. I'm Labour, but let me tell you, those young bastards are something else. Take their orders from Moscow and that's where they ought to go back, in my opinion.'

They talked to the man for half an hour, then excused themselves.

'Have to draw the line somewhere,' he said as they parted, 'and most folk reckon the commies have reached it. Only thing worries me is, what *can* we do? No one in their right mind wants a war, do they?'

'You know,' Nancy remarked over dinner at the Italian restaurant, 'that funny little man said it all. What does happen next?' As she spoke she realized that Tom possibly knew. 'Don't tell me anything you ought not, but I do like to know your thoughts and what's worrying you.'

'There is one guy on whom a lot could depend,' Peterson admitted, snapping a bread stick in half and munching it reflectively. He explained the orthodoxy of Howard Smith's military education, combined with his streak of stubborn ambition.

'Pretty much the opposite to how you grew up,' she observed. 'Sure you're not prejudiced? I wouldn't call you the most flexible man I ever met.' She touched his hand, seeing he was bridling. 'Don't take that the wrong way, sweetheart. It's just that when you go into that "Be clear, let's lay it on the line" routine you don't exactly encourage other points of view.'

'Surely I never spoke like that to you?' Peterson had quite consciously compartmentalized both his life and his language since their marriage.

'You did not and you'd better not!' She laughed. 'But I've overheard certain phrases.'

'Well,' he conceded, 'maybe this man and I do both think our own way's best. The difference is, I'm sending him – ' He cut short the sentence, embarrassed at having revealed too much.

Inevitably she picked up the implication. 'Well, if you're not going wherever it is, that's a mighty relief.' She held his hand tightly across the table. 'Tom, have you thought at all about quitting when this is over?'

'Leaving the Green Machine? Right now they'd never let me go.' He hesitated, not wishing to disappoint her, then decided he must be frank. 'Nancy, darling, I promise to think about it.' He squeezed her hand in return and grinned. 'Be clear, Mrs Peterson, I will give that deal full consideration. When this is over.' Hell, he thought, that's one question which can definitely wait. He had enough problems on his hands for the time being.

The telex from Munich was simple enough. Erich Braun read it once before the coil of paper jumped to accept a further incoming message. As soon as the machine was still again, he carefully tore off the strip bearing his own instruction. One of his office colleagues saw it over his shoulder and made a sardonic comment.

'Pretty exciting copy that'll make for your magazine.'

'They're crazy about those sort of details,' he replied, forcing a laugh. 'What the hell so long as they pay?'

'True, it's all money.' The colleague detached his own telex and returned to his desk.

Braun followed his example, sat down and studied the telex again.

> REQUIRE FURTHER THREE EXAMPLES NATO STANDARDIZATION DEADLINE THURSDAY.

It was signed Diederichs, Foreign Editor.

The standardization of equipment among the thirteen NATO nations had been a subject of study by a NATO Military Agency since 1951, and every defence-oriented journalist in Brussels found himself obliged to write about it eventually. For Erich Braun, however, standardization had a totally different significance. His employers had adopted it and various other commonplace NATO expressions as a simple code for emergencies. This telex called him to a contact meeting at three on Thursday afternoon. Today was Tuesday. It didn't allow him much time. Erich swore to himself, then lifted the telephone and rang the NATO Press Service. He could at least get an ostensible answer to the query telexed back to Munich. He was meticulous about maintaining his cover.

Sysselmann Prebensen had been steeling himself for the interview. The strain of maintaining a façade of cooperation with the Russians was showing in his face after these three months, the lines around his eyes were etched deeper, the long Viking jaw, never fleshy, was now gaunt. He was suffering a curious stiffness in his fingers, which the doctor at the small hospital was unable to explain, except as some kind of psychosomatic response. Dealing with General Stolypin had been Kafka-esque in its distortions of reality, infinitely wearying. Far from being the plain soldier his rank suggested, Stolypin had proved devious and unpredictable. However, the instructions from Oslo were firm: Prebensen had to tackle the General head on.

They met formally in the conference room. Once again the miniature natural flags decorated the centre of the long black table, once again the Russians elected to sit with their backs to the windows overlooking the fjord, which meant with their backs to the light. When they first met here, the outside world had been dark; now the landscape was in strong sunlight, the distant peaks glowing white against a blue sky, the thick ice in the fjord irregular and humped where it was starting to break up. By mid-June there would be a clear passage for ships to dock at Longyearbyen's wooden wharf. Today was 21 May.

Stolypin shook hands formally, then took his seat, Makarov flanking him to his right, the interpreter on his left. The General was, as usual, in civilian clothes of noticeably better cut than Makarov's: the military attaché in London obtained his suits for him. One of Stolypin's greatest talents lay in manipulating the system to his own advantage.

'So, Comrade Governor,' he began cheerfully. 'What have we to discuss today?' Both the question and the attitude were hugely insincere, since his staff's interception of the telegraph traffic from Norway had prepared him adequately in advance.

'My government has instructed me,' Prebensen intoned with all the gravity he could muster, 'to demand a full and sufficient explanation of the construction work being carried out on the Plata Berget mountain and further to insist on investigation of the accident there yesterday by qualified Norwegian government inspectors.'

Stolypin nodded fractionally, drew a silver case from his pocket and methodically lit a thin Russian cigarette. 'Please continue,' he said in the tone of a schoolmaster who is prepared, just prepared, to put up with a pupil's complaints about the food.

'Further,' said Prebensen, feeling the fingers of his left hand tighten in a muscular spasm, 'my government challenges with all the force of its authority the Soviet Union's attempt to install military radar equipment on the mountain in flagrant contravention of Article Nine of the Svalbard Treaty.' He gave great emphasis to the last sentence. This was the moment of truth in the artificial relationship between him and Stolypin, between Norway and Russia. Article Nine prohibited the use of Svalbard for warlike purposes.

Makarov listened, appreciating the significance of the showdown, yet distracted by related thoughts. His body ached as if he had been beaten, and he was exhausted from his efforts both on the mountaintop and in organizing rescue of the injured. Worse, he feared Stolypin was going to pin the blame for the disaster on him, whereas if anyone was responsible it was the dead helicopter pilot.

The interpreter finished. Stolypin drew on his cigarette, careful not to blow smoke across the table.

'Comrade Governor,' he replied quietly. 'I shall take your points in order. I categorically dispute that the accident to the aircraft is the affair of your government. The helicopter was Soviet registered, it crashed on soil under the joint sovereignty of my government.' He looked straight at Prebensen, his eyes hard. 'Had the Royal Norwegian government displayed a more cooperative attitude over the Hopen Island incident, there might be a precedent for collaboration. As it is, we shall make such information as is necessary for the improvement of air safety available to you in due course.'

Prebensen took a deep breath and contained his anger. He felt nauseous and had to conceal his left hand out of sight because the fingers were as stiff as a bird's claw.

'As to your suggestions about the radar – ' Stolypin let his tone relax. He had delivered his first snub, now his aim was to confuse before being insulting again. '– may I in turn refer you to the provisions of the treaty?' He picked up a copy of the short document he had ready. 'Article Four states "Owners of landed property shall always be at liberty to establish and use for their own purposes wireless telegraphy installations. . . ." I am sure you are familiar with the wording, Comrade Governor.'

'The Plata Berget is not Soviet property.'

'The paragraph does not specify that the installations must be on the property. What it does establish is freedom to communicate on private business with fixed or moving wireless stations, whether on board ships or aircraft. Can you deny that there are no existing radar facilities at Longyearbyen with which to assist the safe navigation of aircraft? How can the approach radar we have near Barentsburg serve Aeroflot planes landing 35 kilometres away?'

'That is no justification for the erection of military radar.'

'*If* you had chosen to ask us for information – ' Stolypin let the scorn burn in his voice so that the Norwegian could not mistake it, no matter how the interpreter spoke '– instead of depending on inaccurate intelligence from American space satellites, you might understand the situation correctly.' He let this sink in, estimated that Prebensen was about to rise and walk out, then smiled. 'Would your government seriously oppose the provision of approach control radar at Longyearbyen?'

Prebensen was caught. 'What is your exact suggestion?' he asked.

'To make a compromise and transfer one of the existing radar vehicles from Barentsburg to the Plata Berget.'

'I will put the matter to my government.'

'Your government has already accepted our radar at Barentsburg. We are not prepared to delay. You can inform Oslo that the transfer will be made tomorrow.' Stolypin rose to his feet. 'You will excuse us. We have much to do.' He walked out, leaving Prebensen to wrestle with his conscience and the facts of power.

In the afternoon, while Makarov organized the dispatch of a replacement Hook helicopter from Murmansk to remove the radar vehicle from the landing ground on the small headland of Heerodden near Barentsburg, Stolypin summoned Yakushkin to his office.

The young engineer was still in a state of shock from the crash. Stolypin welcomed him solicitously, had coffee served and questioned him briefly about the transfer of the radar. It was essential to be seen by the West to recover instantly from the disaster. He had assured his mentors in the Politburo that a surveillance radar would be operating on the Plata Berget by tomorrow and a radar there was going to be.

'Well, Comrade General,' said Yakushkin, nervous and afraid of displeasing him, 'the vehicle mounted radar does not have the same range. Only 200 nautical miles maximum against 400 . . . but it is self-contained.'

'Can it be linked to the control centre in the same way?'

'Oh yes. It has its own generator.'

'Excellent.' Stolypin beamed. 'We are lucky to have a man of your technical expertise available. With your help we shall overcome yesterday's tragedy.'

'I will do my best, Comrade General.'

'Believe me, Pyotr Ivanovich. I never doubted you would.' The confidence and reassurance in the words overwhelmed Yakushkin: overwhelmed him and left him vulnerable. 'There will be an inquiry into the accident, of course,' Stolypin went on benignly. 'I shall be asking you to give expert evidence.'

Yakushkin nodded, lulled by the friendliness.

'When that moment comes,' said Stolypin, a shade more toughly, 'you must put all personal scruples aside. The truth is the important thing. Blame must fall where it is due.' He leaned forward across the table with the coffee cups. 'Never forget, Comrade, it is the duty of every Soviet citizen to denounce crime.'

He paused to let the idea take root and saw from the fear in the young man's face that it had. Denunciation was more than a duty, it had been a legal obligation since the Law Code of 1649, when the czars made failure to denounce treason punishable by death. The revolution had not altered this peculiarly Russian concept. It was with this crime that the dissident Anatoly Kuznetsov was charged by the KGB in 1963, because he had not reported a tirade against the Party by some of his students.

Yakushkin went white. Too late he realized where this intimate conversation was heading. 'Yes, Comrade General,' he whispered.

'If incompetence or incorrect decisions caused yesterday's loss of life,' Stolypin concluded, 'you must say so fearlessly.'

Oh God, Yakushkin thought as he left, they want a scapegoat and they won't make do with the dead. He worked the rest of the afternoon in a daze, wondering if he dared warn Makarov.

The next day the vehicle-mounted radar was successfully lifted up to the Plata Berget. When Makarov returned, weary but relieved, he found a letter from his wife.

'My darling Viki,' she wrote in her carefully formed cyrillic script. 'The summer will soon be here and I had longed for word that you had arranged my visit. But now the department says no holidays can be permitted during the present confrontation with the West. Every day we hear on the radio of new imperialist manoeuvres. No one can doubt they might attack at any moment and we are busy with preparation for the defence of the homeland.'

Makarov paused in his reading. All Russians were united in a passionate love of their motherland, a patriotic emotion which, in his few travels, he had never seen elsewhere. Twenty million had died defending Mother Russia against the Nazis. He could well imagine what Svetlana meant by 'preparation'. Air-raid shelters would be provisioned, reservists recalled to their regiments, gas masks issued to the more important civilians, resistance cells activated in factories and on farms. Yesterday's newspapers, flown in from Moscow, had carried banner headlines appealing to the spirit of Stalingrad and virulently denouncing the United States as imperialist aggressors. None the less he was enough of a realist to appreciate that the entire crisis hinged on what he was doing in Svalbard. Perhaps his wife did too because she concluded by saying, 'I know in my heart you are in the front line. Come back safely!'

The long entrance lobby of the NATO headquarters was garlanded with white-tasselled cords and flags. Outside, the limousines bringing the defence ministers of the member countries swung past the huge bronze star and disgorged their dark-suited, grave-faced passengers. Guards saluted. All the normal panoply was in evidence for the Defence Planning Committee.

But the meeting was not merely routine. Even in the press area, with its photographs of recent exercises on display and discarded coffee cups littering the low tables, there was a worried, serious atmosphere. Erich Braun sensed it the moment he walked in. Usually the journalists, hanging about waiting for press conferences, making appointments for briefings by the delegations, hastening with scribbled stories to telex machines and telephones, went about their business as though it mattered intensely to elicit the latest nuance in the Alliance's policy, if only because their own reputations as reporters hung on it. Today things were different. Erich picked up a press release, headed with the blue NATO emblem.

'The May ministerial meeting, postponed two weeks from its original date, will consider urgent questions facing the Alliance. The ministers . . .'

Such a postponement was exceptional. The NATO nations must be desperate to achieve a united front. Erich began moving among his press acquaintances, throwing in a fragment of gossip here, extracting an item there. Almost every foreign reporter had a personal contact among the diplomats representing his own country, each would be able to gauge how his own nation's ministers had voted. Erich had no such relationship, but over the two days he expected to piece together a fair assessment of what had happened in the supposed secrecy of Conference Room 16.

That afternoon, in the International Military Staff wing of the building, Admiral King was closeted with General Anderson. He had flown from Washington in an aircraft rivalling the Presidential Air Force One in the sophistication of the communications facilities on board. Having reached a decision, the Chiefs of Staff did not want the officer charged with implementing it out of touch during a whistle-stop tour of Europe.

'So, Hiram, are we recommending military action be taken, or informing the ministers that we're taking it?' King asked. He was wary of telling the politicians anything, even in closed session when all but a few advisers were excluded.

Anderson considered this. It was a point he had thought through with great care and still not resolved. On the one hand it would be desirable for the ministers to approve, because then the whole Alliance would be involved. On the other, he shared John King's fears about security.

'I guess we have to make the usual damn compromise. Tell them that military action must be prepared for, without specifying too clearly what it is, and ask them to approve the call-up of reserves.'

'I'd go further than that,' said King. 'I'd advocate a naval blockade. I think the Norwegian Prime Minister would agree. He knows as well as any of us that the main radar masts the Soviets are obviously going to erect on the Isfjord will have to be shipped in. We should warn them we're going to stop and search vessels that may be carrying war materials.'

'The Iceland ministers will have something to say on that,' said Anderson sharply. There were several communists in the Icelandic government. Furthermore Iceland was sensitive about its territorial waters, as NATO had cause to remember from the bitter 'cod war' with Britain of 1972.

'The ice is pretty far north now. We could patrol right close in to Svalbard. Way out of Icelandic waters.'

'I don't think they'll buy it,' Anderson growled. 'At least not until after the Security Council debate.' He glanced at the clock. 'We'll know soon enough, anyway. More coffee?' He spoke through the intercom on his desk. 'Hey, Sandra, bring us some more coffee and juice, will you?'

An hour later, the two Americans found that their optimism had not been justified. In the unrestricted session this morning the ministers had made resounding speeches designed to impress the Russians with their singleness of purpose. Now, grouped in the smaller conference room number 1, with translators absent and the requirement for firm policy commitments emphasized by the Secretary-General in his opening remarks, the mood was more sombre. Hardly a minister there did not have lurking at the back of his mind some fear. The Dutchman, normally forthright, had left Holland in the grip of a dock strike which was already dislocating exports and could severely hamper reinforcement of the British Army of the Rhine in Germany. The Canadian and Danish ministers still both preferred the intervention of a United Nations force. Even Sinclair, the British minister, appeared only lukewarm. Only the Turks, the NATO country with the largest army in Europe, accustomed to their Armenian border with the Soviet Union, sided wholeheartedly with the American Secretary for Defense in supporting military action. The upshot was agreement on minor increases in force readiness and a request that the military staffs prepare plans for a partial naval blockade.

'From this moment on,' the American Defense Secretary told King and Anderson afterwards, 'any meaningful discussion will be strictly bilateral with the British. And the Norwegians, if they come in.'

'I'd certainly be happier that way,' King said. 'The security aspect worries me.' It had occurred to him that the British minister's attitude might simply be calculated to put the Russians off the scent, assuming they acquired reasonably accurate information about the meeting. Too many spies had been uncovered in NATO for a prudent man not to assume there were more.

'We'll push for a blockade when the foreign ministers meet tomorrow. That's all. Thank you, gentlemen.'

When the carefully worded final communiqué was issued next evening, Erich guessed that the foreign ministers had also been divided.

'The North Atlantic Council met in Ministerial Session in Brussels,' it began. 'Ministers renewed their faith in the North Atlantic Treaty, which guarantees the preservation of peace and international stability in accordance with Article Fifty-one of the United Nations Charter. . . .' Erich smiled at the careful invoking of the world body and skipped through the next paragraphs. 'Reviewing the serious implications of the recent Soviet action in Svalbard, ministers reaffirmed their commitment under Article Four of the Treaty to consult whenever the territorial integrity of one of the Parties is threatened . . . they considered that the present Soviet action constitutes a threat and further considered what action should be taken to maintain the security of the North Atlantic area. Certain precautionary military measures have been ordered . . . the Allies will remain in close consultation.'

'Blah, blah, blah,' Erich murmured to himself and took an extra copy to show Sandra. It would be useful ammunition and this coming weekend was the one long earmarked for their flying excursion to Deauville, the time when he must bring her to the moment of truth. The Stasi could wait no longer and he understood why. Yet in a rare moment of stubbornness, he had insisted to his contact on playing Sandra his way. 'I must get her away from the crisis atmosphere of Brussels, give her some excitement,' he had reiterated three days ago. 'Anyway, don't you want those photographs?' Much planning had already gone into the Deauville weekend, which increasingly felt like a last fling before war enveloped Europe.

He was standing in the Press Centre concourse, the communiqué and other papers in his hand, wondering whether to call Sandra on the internal telephone, when he saw her colleague Julia coming towards him. The concourse was flanked by a bookshop and post office, while at the end were glass doors to a canteen used both by NATO staff and journalists. Julia was clearly on her way there. Though he had only met her once, out at Grimbergen, he recognized her instantly. She was not very tall and walked with her head up, long blond hair rustling her shoulders, hips swaying under a tight skirt, breasts thrusting against the restraint of a blue silk blouse, the gold bracelet flashing on her ankle. She looked only a fraction less brassy than a very high-class tart. Every cowboy's dream, he thought, cynicism and lust balancing each other in his mind.

'Hallo there!' She had stopped. 'Fancy seeing you again.' She looked him straight in the eye, half smiling provocatively.

'I do represent a magazine.' He tried to sound offended.

'Of course, I was only teasing.' She let the smile develop,

delighted at what chance was delivering. He was one hell of an attractive man. 'Well, are you going to invite me for a drink, or do I have to stand here until Sandra comes?'

'Oh, no.' Her directness caught him off balance. 'That will be a pleasure.'

To his surprise, Julia let him precede her into the restaurant, as is the German custom. It was an intimation that she was adaptable as well as self-possessed. Holding her chair for her, he felt unexpectedly disturbed that she must be at least three years his senior. Courteously, he departed to fetch her a gin and tonic.

Julia watched him thread his way among the tables to the bar. 'He's even better than I remembered,' she murmured to herself. 'Darling Sandra *has* caught a handsome fish.' She had marked him down out at the Grimbergen flying club and assumed their affair would not last long. Yet here he was, six months later, as large as life and twice as sexy and still around. Oh God, but she could use a man tonight. Her insides began to turn in an inexorable, familiar way and she felt herself flushing. She knew Sandra was dining with the Brigadier's family. Would anyone ever find out? Then, as her eyes followed Erich coming back with the drinks, she found herself trembling with a different emotion. Sandra must be a prodigy in bed to keep him happy. Was it possible? She had joked about Sandra's Kraut lover, but who the hell was Erich? She decided to find out. When he sat down again she was completely composed.

'Just exactly what I needed!' She sipped the gin. 'You've saved my life. What a day!' She winked across the top of the glass. 'So, which of our little secrets have you been telling the world about this time? You've all been writing on Svalbard, I suppose?'

'Trying to. It is not so easy.' He pulled the communiqué from his folder, still feeling ill at ease, his language stiffening accordingly. 'What kind of a story can anyone make from that, I ask?'

Julia wrinkled her heavily powdered nose. 'But surely that just means what it says? They won't do a thing.' To her, putting reporters off the scent was an automatic reflex.

'You think not?' He was afraid Sandra might come in, yet could not resist such an opportunity to crosscheck whatever she told him tomorrow. He forced himself to relax, attempting to match Julia's mood as she chattered about social life in Brussels.

A quarter of an hour later she asked slyly, 'Do you think Sandra is coming?'

'She must have forgotten.' He smiled in a way which confirmed what she clearly wanted to be told.

'Then let's get out of here. I've had enough of the great Alliance for today.' She decided to continue jumping in head first. It usually paid. 'I must go home for a bath, though. Where shall we meet?'

They made a show of saying goodbye in the canteen. When Erich eventually left her apartment in the city at two the next morning he was amazed and exhausted, yet not entirely content. He had learned several things about Julia and very little about NATO. He had also taken an absurd risk, one which he drummed into himself he could not repeat. Six months' cultivation of Sandra would be blown if she found out. She would go mad with jealousy. But, damn it, he finally admitted as he clambered into the hard bed in his rented room, if living was basically a series of experiences, then he had added to his life in these few hours. What a woman! The trick, he decided as he fell asleep, was to compartmentalize all these experiences, so that one never trespassed on another, so that the pleasures and the perils were kept safely separate.

For Tom and Nancy Peterson the period of enforced idleness while the Norwegians explored every last diplomatic initiative turned into an unexpected second honeymoon.

Then came the day of the NATO meeting. That evening one of the Navy duty officers rang and said Admiral King would be across and wanted to see Tom first thing in the morning.

'The President personally authorized the raid last night. I can tell you something, Tom, it's been one hell of a business getting this act together. I had to convince the Norwegian Prime Minister myself. That trawler the British converted will take your boys in and there will be a Norwegian officer on board.' He gave Peterson a sharp glance. 'Wish you were going to be on the operation yourself?'

'I do, sir.'

'Comes a time when we all have to give up leading from the front. Someone has to run the operations room for a raid like this.'

Peterson chewed his lip. From the Norwegian headquarters outside Bodo he would have special radio links to the team, all the available intelligence on Soviet movements, satellite pictures, communications intercepts. But the week allowed for the trawler to plough its way from the North Sea to Svalbard was going to be very long and he was going to have one very fretful wife left by herself in Kensington.

'The guts of this mission are as simple as ever,' King concluded, 'to show the Soviets we are prepared to destroy military installations in Svalbard if necessary and back it afterwards with a call on the hot line making clear it was a strictly limited action. Well, Peterson, I'm not coming to see them off. That would be somewhat conspicuous. But you tell them I'll be with them all the way.'

'I will, sir.'

As usual, once Admiral King was satisfied a project was in good hands, he wasted no time interfering further. 'You have any problems, contact me direct. I'll be available.'

'Thank you, sir.' Peterson snapped confidently to attention as the Admiral left the room. But for the first time in his career as a soldier he began to understand the full loneliness of command, when you consign others to battle and can help them no further.

6

The grass airfield, twenty minutes' drive from Brussels, was too pretty to be true, Sandra decided. Where real airports have ugly concrete buildings, No Waiting signs and officious police, Grimbergen had a traditional Flemish house with a steeply pitched roof as a restaurant. Outside there were tables and chairs on the grass a few yards from the diminutive white control tower. Small aeroplanes, painted in vivid colours, were parked like toys and every now and then one flew overhead in the blue sky, circling lazily to land.

The whole scene was like a Dufy painting of summer, with planes instead of yachts. A man in a check shirt cycled past and waved as Erich carried their case out to the Cessna. She waved back, divinely happy. Life was so very good, the worries of work and the threats of war banished for a whole weekend.

The Cessna had four seats, high wings and a single engine. It was white with an elegant blue stripe along the body. Sandra sat in the right-hand seat, and took over the dual controls as soon as they were on course, smiling nervously as Erich made occasional comments on her flying. As they passed over the drab countryside near Lille, factory chimneys pouring smoke in the distance, she realized with a flicker of horrified curiosity that the long zigzag indentations and grassed-over craters in the land below were the battlefields of the First World War, the trenches and the shell holes where so many died. The thought of Erich's and her own forbears locked in such bloody conflict disturbed her and she was glad when they were over the orchards of Normandy. The winding river Seine came in sight, huge barges chugging ponderously down it, and they crossed the estuary over the wide span of the new Tancarville bridge and landed at Deauville.

The happy Dufy impression returned as soon as they were in the resort, its wide beach decorated with flags and bathing huts.

'It's perfect, darling,' she breathed as they were shown their rooms overlooking the sea. The moment the maid was gone, she put her arms round Erich's neck and kissed him lovingly.

He responded passionately, undressing her, then drew her down on the crisp counterpane of the big bed. He had clear orders as to what he was to do this weekend. The man who had occupied the room until this morning had made some minor adjustments to the wiring of the service bell. When pressed it also actuated a tiny camera hidden high in the curtain pelmet and focused on the bed. Personally Erich considered this absurd, typically Russian, in fact. His own way of working on Sandra would be far more effective than blackmail with indecent photographs. Certainly Julia would tell anyone who tried such an old trick to go to hell. However, he did as he was instructed.

'Champagne,' he murmured when they had finished making love. 'Now we must have champagne.' His contact had specifically authorized the expense.

As he pressed the bell Sandra was still lying with her legs apart on the rumpled bedcover, smiling in abandoned contentment. The sunlight streaming in through the window reflected off the wall and bathed her thighs and breasts in a diffused glow. He touched the button twice more, making sure his own nakedness was also visible. He took a hasty shower and had his trousers on again when the waiter arrived.

In the afternoon they ventured into the sea. Though the water was too cold to be enjoyable, he swam briskly to work off the alcohol. He felt the cloying adoration of her remarks needed washing away too. She seemed to see everything in terms of a woman's magazine romance.

Over dinner in the restaurant at the Casino he held her hand and proposed: cautiously, as befitted a struggling writer who was splurging a month's earnings on one glorious fling. She gave him the cue as the filet mignon was brought to the table and with it a bottle of Beaujolais Villages.

'Erich, darling, this must be costing you a fortune! Are you sure I'm worth it?'

The note of realism surprised him, but he instantly exploited it, reached out to hold her fingertips. 'Yes, my love. You're worth every penny. I'm a very lucky man.' He gazed into her eyes with deep seriousness. 'Sandra, darling, I can't wait any longer to ask you. You know I'm in love with you. Will you marry me?' He spread both hands out on the table. 'Darling, by the autumn I'll have the new job. In Berlin we can live near the Wannsee, where there are woods and flowers. You'll like it there, I promise. Berlin is a marvellous city. Sandra, will you?'

'Oh, Erich.' Tears were filling her eyes, she could hardly speak. 'I love you so much too.'

Their hand-holding was interrupted by a low cough from the waiter. 'Broccoli, madame?'

'What a moment for vegetables!' Erich laughed, tickled by the absurdity of the whole scene. 'Yes, broccoli.'

'Bother all waiters,' Sandra said. She didn't find it funny, but it gave her a chance to think. 'Of course I have to give notice. Three months it's supposed to be.'

'Then as soon as I have the job we become officially engaged. At Christmas we marry.' He smiled with the authority of a man who knew how to arrange his affairs. 'How about that?'

'You must come and meet Mummy first.' Plans were

running in her mind like quicksilver dreams, a wedding dress, bridesmaids in the parish church, photographers. What would her mother think of his being a German?

Erich saw something of these visions in her eyes. 'She won't mind? My being a foreigner, I mean.'

'Oh, no.' She was less certain than she sounded. 'As long as she knows you love me, she'll be thrilled.'

For Sandra the rest of the weekend was an excited, delirious dream. On Sunday, they paraded along the shops, gazing in the windows of Paris jewellers until Sandra spotted the perfect ring, diamonds alternating with blue sapphires in a tiny chequerboard band. Erich encouraged her with uninhibited enthusiasm, noting the telephone number so that he could ring from Brussels and inquire the price. 'Too bad they're closed today,' he kept saying, careful to keep any sign of relief from his voice.

The weather was clouding over when they landed back at Grimbergen in the early evening. As they drove into the city a light rain began to fall, the mournful drizzle of many Brussels days. On the way to Sandra's apartment Erich called at his office, let himself in quickly to the deserted building and re-emerged holding a telex message, his face set and anxious. He let her read it in the fading light.

> REQUIRE FULL INTERPRETATIVE BACKGROUNDER
> SVALBARD CONFRONTATION SOONEST. YOUR
> LAST INADEQUATE MANY RESPECTS NOTABLY
> NATO ATTITUDES. DIEDERICHS.

She knew at once what he was going to ask. The sunlit glory of the weekend began to fade, her mood became as dank as the rain enveloping the city. At the apartment, she tried desperately to cheer him up, fumbling in her hurry to extract ice from the freezer for his whisky, asking too anxiously what he would like to eat.

'How can I think of food when I receive such a reprimand?' He paced the sitting room as if in a frenzy until he knocked into the galleried trolley. Then, creating a pantomime of despair, he slumped into the depths of the sofa. 'You have to help me. This time more than ever. My whole future is at stake.'

'Erich, darling.' She sat beside him, trying to hold his hand. 'Even wives are not allowed to tell their husbands.'

'Nonsense, of course they talk. Don't you trust me?' He switched to being affectionate and squeezed her hand in return. 'My darling, I am not going to put secrets in my article. Didn't you read the telex? "Interpretative" it said. What they want is analysis. But it must be based on knowledge, not speculation.' He kissed her cheek. 'My darling, I am not going to spill your secrets to the world, I promise.'

She clung to his hand, tormented. 'I'll see if I can help. But really, we only deal with military plans, not the politics. I don't know very much myself.'

'But plans are made according to policy. You can find out.'

'Perhaps.' She was terrified of hurting him. 'Darling, I'm exhausted. It's late. Why don't we just go to bed?'

'Yes. I have to start early,' he said abruptly.

Inevitably, once they were naked and between the sheets, he began to feel her body. She responded with a mixture of relief and passion. The act of sex would make everything right again. They never quarrelled in bed.

He was inside her, her legs tight around him, when she felt him convulse as if stabbed. 'No,' he cried out, thrusting himself away from her, rolling off the bed. 'No, it isn't any good.'

'Darling, come back, please come back.' He had left her at the moment of climax. She couldn't bear being left.

'If you don't trust me we can never marry.'

She could discern his outline reaching for his clothes, the ghost of a shirt being pulled over his head.

'What are you doing, Erich?' It was all she could do to keep her voice down. 'What do you mean?'

'I am going to my place. If we have no trust then there is nothing. Null, null, null.'

'Erich!' she sat up, still panting, and seized his arm. 'Don't go, you mustn't, I can't bear it.' She felt his tenseness yield. 'Darling, I will find out, I will, I will. Don't leave me, darling, please don't go.' All those dreams, the wedding, the baby she so desperately wanted before it was too late, the tremendous agony of losing her great love, all boiled inside her. 'I do trust you, Erich, I promise I do.'

She felt his hand round her shoulder, his lips brushed her cheek and she began to tremble and cry. Very gently he guided her back into the centre of the bed and began making love again. But he paused to slip off his shirt first. He didn't want it creased in the morning.

'God, what a waste!'

Paul Mydland glanced at the English major, dressed now in a thick grey Norwegian sweater and dirty old trousers, peering at the sea through the spray-slimed windows of the *Northern Light*'s wheelhouse. Mydland was in a seaman's short coat, the soiled property of a genuine trawler captain, and his chin bristled with stubble. He was letting his beard grow as an elementary disguise.

'What is a waste? I don't follow you.'

Simon Weston chuckled, enjoying exaggerating his own Englishness. 'Haven't you noticed the time, old boy? Midnight of Saturday. We could be doing our thing with all those beautiful girls of yours. Instead of which we're trundling around the bloody North Sea waiting for a bunch of Yanks. What a waste. Doesn't matter which country they belong to, pongos never get their priorities right.' He shook his head despairingly, then remembered that his companion was an army officer. 'Sorry, old chap, didn't mean to be rude. But why not Sunday night, eh? We can't land up there until more ice melts, anyway.'

They were ten miles from the Norwegian coast, near the island of Loppa, north of Tromso, having emerged from the protected waterways among the islands during the evening.

Loppa was far up in the Arctic Circle, almost exactly on the 70 degree parallel of latitude. Occasionally the clouds parted to reveal the midnight sun, an orange ball hanging low over the northern horizon, but its diffused light lay all the time on the dark water, ruffled only by a light wind.

'Really hurts me,' Weston insisted. 'I fly in from Scotland, pick you and the trawler up and never even stop.'

Mydland laughed. He knew a little about the commandos' nights out at the SAS hotel in his home town of Tromso, where now the snow had melted, leaving only the mountaintops patchily white, and where crocuses and daffodils were pushing up through newly thick grass. Normally at this time of year tourists thronged the streets, buying sealskin souvenirs and feasting on fresh salmon in the fish restaurants. But this had been a season of abnormality. The Svalbard sailings, due to start in a fortnight, were closed to foreign passengers. Mydland doubted whether they would sail at all. Over the past weeks he had not dared to contact Annie, though he had received a message through a mutual friend to say she was well. Nor had any word of what she and Folvik were reporting reached him. He had been as cut off from the military as if he had only been briefly recalled for annual training, and yet the mining company had been strangely slow in finding him a new assignment. Then had come the call from the North Norway Defence Command at Bodo and here he was, his country's representative on a raid which he could scarcely believe was taking place.

Yet both he and Weston had heard the unscheduled talk by the Norwegian Prime Minister on the ship's radio. 'Norwegians, compatriots,' the old man had begun with rare emotion, 'we have been thrust against our will into a time of crisis. We have tried to reconvene the Paris Conference which established the Norwegian Crown's sovereignty over Svalbard. But both the Soviet Union and India have refused to sit at the conference table with the representative of South Africa.' He had paused, his voice unsteady for a moment. 'Many of you will sympathize with that attitude.

Unfortunately it has closed the obvious door to a settlement.

'Despite our protests,' the Prime Minister had continued, 'the Russians have now positioned military radar on Svalbard in flagrant breach of the Treaty. They have vetoed our latest attempt to obtain intervention by the United Nations. We must now take action in cooperation with our NATO Allies.'

Mydland had pictured the agony of mind this must have cost a government so pledged to peaceful means of settling disputes. The Prime Minister had ended with an appeal to the population to remain calm. Suddenly it had occurred to Mydland that this leader he had never met must have personally authorized his own part in this mission. That had given him more cause for thought than the speech.

The *Northern Light*'s diesels throbbed gently, just giving her headway against the tide, the steadiness of her passage as unreal as a dream. Things don't happen like this, thought Mydland. The whole of his life seemed in suspension as they waited for the American destroyer to creep in from the west and disgorge the Delta Force team. The smoke from Weston's pipe curled round the unlit warmth of the wheelhouse, and the little radar screen shone dull yellow as its arm of light rotated monotonously round and round, each sweeping circuit making various dots glow afresh. Three of those tiny blobs were Norwegian patrol craft, sealing off the approaches from intruders. Others must be fishing boats. Soon a new one would edge in from the side of the screen and that would be the American ship.

'What d'you think of this expedition, Paul? Riskier for you than most of us, isn't it?'

'When the local people helped an operation like this in the Second World War,' he observed. 'the Nazis executed hostages.' He had a sudden vision of Folvik, round-faced and bewildered, his wispy hair in disorder, being stood against a wall by Russian soldiers. And Annie?

Weston drew on his pipe, thinking. They were neither of

them born then, their mothers were schoolgirls. The raids on the Lofoten islands, not so far south of here, were fading into history now, seldom recounted by survivors as personal experience. The burning of threequarters of the buildings in Finnmark by the Nazis was history too, but also a folk memory locally. It was easy for the British and Americans to forget what they had never experienced. For nations which had been invaded it was less easy. Weston's hobby was military history, one day he intended to write a book about commando operations, and he felt himself at the divide between the reading of history and the making of it.

'Well,' he said, exhaling smoke, then rubbing his palm across the steamed-up window. 'I don't imagine the Russians would take reprisals over this, it is potentially so public. They can get away with using chemical weapons in Afghanistan, or Eritrea. Nobody gives a damn about a lot of wogs.' He grinned. 'Especially if they're killed on behalf of other wogs. No, as I see it this is one hell of a bloody great bluff.'

'Bluff!' Mydland had been told that the United States was determined to remove the radar and make the Russians back down.

'Essentially, yes.' Weston was in a ruminative mood, and spoke as though considering some abstract proposition. 'Just as the Russians taking over Svalbard was bluff. No one wants another world war, agreed?'

'One hundred per cent.'

'Then if either superpower wants to improve its position militarily it will go only as far as it believes it can *without* a war.'

'I agree.' His own experience with Stolypin suggested to Mydland that the Soviet General had been treading warily.

'So our task is to demonstrate that the Russians have now gone too far.' Weston paused, the pipe smoke eddying in the pale light. 'This raid is telling them that if they pull out the Americans will let them do so without loss of face. I'll bet any money you like that as the radar is blown up some kind of message will be sent on the hot line from Washington to Moscow promising that no one else is going to know. No names, no pack drill. It's a nice little secret between you and us. So long as you find an excuse to pull out. But . . .' He paused again to underline his argument. 'But if we fail to destroy that radar, don't tell me the United States will go to war. Let alone your own country and mine. Not over a godforsaken chunk of the Arctic. Not when the price could be missiles landing on mainland America. That's why I call it bluff.'

Mydland thought about this for a while. The corollary was obvious. If the raid was not a total success they would have risked their lives in vain. He said as much.

'Too damn right.' Weston continued to stare at the sea. 'And if it fails, don't kid yourself. We shall be expendable. Which is what we're employed to be, isn't it?' He let the sudden cynicism fade from his voice. 'There you have the professional soldier's dilemma, eh? If you suspect the plan is faulty, do you say so to save your own skin or do you credit the generals with greater powers of perception than yourself? Very tricky.' He chuckled and fell silent.

Far off a light winked across the water. Weston turned his attention to the radar. A luminous blip was slowly progressing towards the centre of the screen. 'That'll be her,' he said abruptly. He took a signalling lamp, opened the wheelhouse door, letting in the chill air, and flashed a brief reply. He had been ordered to keep strict radio silence.

Twenty minutes later the grey hull of the destroyer was alongside, lines were secured and Peterson's head appeared at the rail. He swarmed down a rope ladder to jump onto the trawler's deck.

'Good to see you.' He shook Weston's hand warmly and turned to the Norwegian. 'You must be Paul Mydland? Glad you're here. Thought we might have a few words while the men are coming aboard. We can't hang around too long.'

As the team lowered packs and equipment, Peterson took Mydland into the wheelhouse and quickly discussed the plan.

'You'll have plenty of time to go over the finer points with Craig Clifford. Now, I know you may have to lead them in. But I ask you to accept his orders at all times, OK? Be clear, there can't be but one boss on this operation. You're a captain, right? Good. Then he's senior anyway.'

'There is one communications question, Colonel. I have a set to link with my own headquarters.'

'Is that necessary?' Peterson was immediately wary. 'Our radio man is pretty good.'

'It is quite usual for a liaison officer to maintain his own communications. I mention it because I must be free to use this link at my own discretion.'

'Hmm.' Peterson was thinking fast. In any normal situation, a liaison officer certainly would do this. But he didn't want to increase the risk of detection. 'Yeah, I suppose so. My orders are for Major Clifford to preserve radio silence except on arrival and before the attack. We have a confirmatory codeword.'

'I understand, sir.' Mydland knew the Americans had not been informed about Folvik, though they must assume that information came out of Longyearbyen. 'I also have my orders.'

'OK.' There was no point in disputing this, it was the penalty of the Norwegians only being brought into the act at the last moment. Yet it blurred his clear-cut vision of the operation. He felt a source of potential dispute had been introduced, more than one of useful intelligence. The results of satellite photography of the Plata Berget would be available to him incomparably faster than anything the Norwegians could produce. The thought reminded him about the ice. He returned to Weston. 'Be advised the wind up there is shifting the ice north. The Isfjord should be clear for navigation in a few days.'

'I don't think we'll be going quite that far.' Weston

smiled. 'I wasn't planning to call at Longyearbyen. We'll be OK if the coast is free of ice.'

'Let's work on that assumption.' Peterson swung himself up the rope ladder, then looked down from the destroyer's side and raised his fist with the thumb outstretched. 'Good luck,' he shouted down. 'Wish I could be with you.' He genuinely did. The operation was his baby and he didn't quite trust others to carry it through.

'Collective responsibility,' Stolypin muttered angrily. 'The fallback of a committee of pensioners!' Though, he admitted to himself, some of those old ones are shrewd. They don't intend being caught out.

He was leaning back in comfort, watching the clouds slide by as the VIP Tupolev droned back towards Longyearbyen and pondering the terms of his new instructions from the Politburo. His recall to Moscow for consultations had been sudden and unexpected, dictated by a trusted agent's report to the KGB that Norway was about to demand an immediate Soviet withdrawal from Svalbard. The peremptoriness of this diplomatic *démarche* followed strangely on the heels of NATO communiqués which were transparently vacillating. Stolypin thought it smelled wrong; the Politburo's fifteen members had been divided as to its significance.

'So, all at once these four million people are going to let their king put on his uniform and lead them into battle,' Kutuzin, the hardliner, had remarked derisively, looking around the green baize table at his colleagues. 'Someone must have given them a shot in the arm.' He had fixed his gaze on the KGB chief. 'Who do you suppose has done that, Comrade General?' Kutuzin had been in a sarcastic mood and Stolypin, flown back to give advice if required, had feared that the lash might be turned on him.

The KGB chief was accustomed to it, though. 'The Americans,' he had answered coolly, looking straight back through those thick glasses, his broad face as smoothly impassive as ever. 'We know it is the Americans.'

Stolypin had listened fascinated as the arguments developed. Then the Politburo members had abruptly realized they had forgotten his presence and the Party Secretary had told him to leave, thanking him for his assistance.

Two hours later, Defence Minister Marshal Ustinov had summoned Stolypin and issued the orders which now, immediately on his arrival back at Longyearbyen, he would start to implement. They left him little to be grateful for, save that if the Americans were rash enough to try a naval blockade the response would be coordinated from Moscow. In a sense that was a major weight off his mind, though it still left him keenly aware that, just as the old in the Politburo would rely on 'collective responsibility' to duck unforeseen consequences of their decisions, so they would equally unhesitatingly utilize him, Stolypin, as a scapegoat.

'We are approaching Svalbard, Comrade General.' The steward disturbed his reverie. Through a gap in the clouds below he caught a sight of the sea and a warship, its wake flecking the water. He strained to watch it, but the curvature of the small window prevented him. Ours or theirs? he speculated, as the cloud closed off his view. The Barents Sea would be the first battlefield.

The Tupolev began its descent, down to those treacherous fog-shrouded mountains. He thought of the radar up there on the Plata Berget charting his progress across the frozen land. It was no thanks to Makarov that the installation was working. He had a surprise in store for Makarov. Were it not for his order of Aleksandr Nevskii, the justice would be far rougher. Within minutes of landing he summoned the Colonel to his office.

Makarov was weary, he had been working long hours during the General's absence. New problems were arising with the construction of the Isfjord radar. However, he squared his shoulders, saluted and made a show of hoping his superior's short trip to Moscow had been fruitful.

Stolypin didn't waste words. 'Certain decisions have been made, Comrade Colonel,' he said, unsmiling. 'One is that Comrade Yakushkin is promoted. From today he takes over the supervision of all radar construction projects in Svalbard.'

For a second Makarov was relieved, then he stiffened as the implications of this sank home. He was being blamed for the disaster on the Plata Berget without the benefit of a court of inquiry, without being allowed to defend himself. What had young Yakushkin accused him of? He looked straight at Stolypin, saw the dislike in the General's eyes, the tiny crease of pleasure at the corners of the thin mouth. So Yakushkin *had* denounced him! It was the only explanation. In the same instant he knew he couldn't blame the boy. Stolypin would have put pressure on him, unrelenting pressure. But what was coming next?

'As you wish, Comrade General,' he replied, keeping a tight hold on his voice.

'You are assigned to other duties,' Stolypin said, angry at eliciting so little reaction, not that he had expected Makarov to become emotional. 'You will be responsible for the local defence of installations in Svalbard.' He underlined the relative ignominy of this task by reciting the troops at the Colonel's disposal. 'You will have one company of the Spetsnaz under command and the patrol detachments of the border police. You will submit a detailed plan to me tomorrow morning.' The sly smile developed. 'Your Angolan experience should be useful. Very few men to cover a huge area. Not, fortunately, that there is any serious likelihood of attack.'

He dismissed the Colonel, well satisfied with his own performance. The accident with the radar and the awkward explaining he himself had been forced into was to some extent avenged. Furthermore, if the Politburo's assessment of NATO's intentions proved faulty, then Makarov's head was laid ready on the chopping block. Like an insurance broker, General Stolypin believed in laying off risks.

*

Every day Folvik used the vantage point of the airport tower to gauge the progress of summer's coming, noting the slow melting of the snow along the shore and the gradual break-up of the ice in the fjord. By May the airfield runway had become completely clear and soon patches of dark ground began to show on the slopes of the Plata Berget, though the mountaintops remained capped with white. The summer fogs became more frequent too, layers of cloud shrouding the whole sky so that looking across the fjord all you could see was a narrow horizontal strip of Arctic landscape beneath a grey pall. He noted the climatic changes because he had done so last year and the year before and he was quite consciously clinging to every tried routine for protection. He was not a fanciful man, nor given to nightmarish imaginings, but he was convinced the social structure of the mining settlement was breaking down around him. The belief was not aided by the Russians' sedulous insistence that everything was friendly and normal. How, for example, could he organize a team to play football at Barentsburg in the first match of the season when two of his best players were talking of blowing up the mine rather than let their coal go to Murmansk?

The cumulative changes of three and a half months frightened Folvik, and not because of his own few attempts to radio intelligence back to Oslo. Strangely, he found he was not fearful for his own skin. He was afraid of open and bloody conflict between the miners and the Soviet border police. The imposition of 250 Russian police, plus many civil servants, represented a one-third increase in Longyearbyen's population. Miners were forced to share rooms: Annie had only kept her tiny flatlet to herself because she was a woman. Supplies in the shop had dwindled as the Russians, hungry for all the luxuries they could never get at home, exploited Stolypin's insistence that roubles must be accepted there. Even their low pay had denuded the shelves of transistors, TV sets and household goods. Since the roubles were not convertible, the shop owner refused to restock. Food was having to be flown up from Tromso in unprecedented quantities, yet fresh fruit, good meat, beer and other items which Norwegians regarded as basic were in short supply. The mining company had been forced to introduce what amounted to rationing.

Perhaps the worst blow, certainly the one which had brought the miners closest to mutiny, concerned the telephone. Since the satellite link had been opened in 1981, the miners had enjoyed long conversations with families and girlfriends on the mainland. Sociologists had noted even before the satellite link that the average length of telephone calls from Svalbard was much greater than from any other part of Norway. But Stolypin, transparently worried about espionage, had stopped the automatic dialling. All calls had to be routed through the exchange, so that they could be taped, translated and the caller identified. Both this surveillance and the resultant delays infuriated everyone.

None the less, the long-planned football match took place on the Sunday after Stolypin's return. Officially it was symbolic of friendship, but it took place at the Russian settlement in case of trouble. The football ground boasted no grass. The white-painted goal posts projected from soil flecked with coal dust, the buildings of Barentsburg in the distance, a plume of smoke rising from an industrial chimney. The rearing snow-covered mountains beyond emphasized the incongruities of the scene. The match was a contest of works teams which could have been taking place in the Ruhr, or in the Midlands of Britain. The crowd of Russians marshalled near the touchlines had the rough, labour-ingrained faces of miners anywhere. The Norwegians of the away team had few supporters, a bare dozen men, tactfully grouped together to avert trouble.

The team had arrived on the Sysselmann's ship, capable of taking cargo and a few passengers and with a helicopter-landing platform mounted prominently over its stern. Folvik reckoned they had been lucky to get to Barentsburg at all. Navigation down the fjord had not been easy. The ice was drifting erratically with the wind, forming a barrier across the far side of the water, a barrier which from deck level looked like a line of breakers, but which was as near solid as made no difference. Fortunately the ice along the southern shore had broken into large hunks, floating miniature icebergs slowly melting into strange jagged shapes that one could imagine were castles, crowns, anchors. Most glowed with an inner turquoise light, the real ice blue. The ship had edged cautiously through them, reaching Barentsburg late.

The settlement was as Folvik remembered it. One long brick building, like a factory, had a slogan atop its roof in letters ten feet high: 'WORK FOR THE COMMUNIST PARTY AND THE PEOPLE'. That sums it up, he thought cynically, and wondered as he always did if they would have imported a few star players to ensure the victory of communism. He also prayed that Sven and Lars-Erik would behave themselves; they had been in a truculent mood all day.

Because they were late, the match began almost immediately. Within minutes the Russians had scored and Folvik's attention was caught by the distinctive cheering which broke out from a small group of Russian spectators in horizontally striped blue and white jerseys. He turned to the chairman of the Russian club, who was escorting him as protocol demanded, and made a remark about the forward's goal.

'He's one of the best,' the Russian observed in heavily accented Norwegian.

By the end of the first half the Russians were ahead by three goals to one and it was obvious that the striker whose friends wore striped jerseys was a very talented player.

'He's not a bloody miner,' Sven said savagely, as they hastily discussed tactics in the interval. 'Look at his hands, look at his face.'

'They've brought him in,' added Lars-Erik. 'I bet any money they have.'

'Now calm down, boys,' said Folvik, his hands fluttering slightly as he spoke. 'This is just a friendly match.'

He could hardly have employed a worse term.

'Friendly!' Sven exploded. 'I can recognize an enemy when I see one, FF, even if you can't.'

The moment Folvik had feared came within minutes. Trying to break through the Norwegian defences, the agile Russian striker hacked Sven's ankle and both men fell. Sven, who was heftily muscled, rolled over in the earth, clutched his ankle momentarily, then stumbled to his feet and slugged the Russian hard in the stomach. Lars-Erik dashed across the field. Several of the Russian team did the same, and then two of the striped-sweater spectators ran on. In seconds there was a mêlée, with the referee's whistle shrilling. To his astonished horror, Folvik saw Sven, whom he regarded as the toughest man in the team, felled with a vicious chop across the side of his head by one of the spectators.

'Hey! Foul!' he shouted.

The Russian chairman looked at him coldly. 'It should teach them to attack a Spetsnaz,' he said, then stopped embarrassed, and lumbered out on the field himself, shouting orders. Before he got there, police had separated the fighting men and were marching them off. But Sven lay inert on the ground until a stretcher was brought.

Folvik was furious. Seeking out the Russian chairman among the officials now clustering the field, he demanded justice.

'Our man has been savagely attacked by a so-called spectator. What did you call him, a Spetsnaz?' He had no idea what the word meant, maybe it was some kind of karate club. Certainly the blow, which had left Sven unconscious, was no ordinary one.

The embarrassment of the Russians at his protest was extreme. A doctor was summoned and when Folvik insisted that the match be suspended they agreed at once. Then, gradually, their apologies became overlaid with stiff formality. They insisted that the scuffle was entirely the Norwegians' fault, denying that foul play by their own striker had started the trouble. The Norwegians refused as one to accept refreshments, and left almost immediately.

Sven died on the voyage back. The team were raging. If they could have returned and killed a few Russians, they would have done so. Folvik stood by the rail, watching the mountains slip by. He was puzzled, because a memory kept tugging at him, a magazine photo of Russians in blue and white striped sweaters. Shortly before they docked at Longyearbyen he remembered what it was. The picture had been of a Red Army parade in Moscow. The soldiers goose-stepping with their rifles held across their chests wore blue berets and open neck tunics with blue and white striped sweaters underneath. The caption announced that they were the élite of the Army: the paratroops.

So what on earth, Folvik asked himself, were paratroops doing in Barentsburg? He could guess why they had worn the sweaters off duty: it was like waving a flag in support of their players. That evening he and Annie transmitted the information to Oslo.

'Bastards,' said Annie softly. She was not given to swearing, but she had liked the bearish Sven. For almost the first time they had sent intelligence that could not be obtained from American satellite photographs, the kind only an observant person on the ground could produce.

'Yes,' Folvik said. 'There is quite a difference between policemen and special troops. At least in terms of the Treaty.' He felt oddly happy in the midst of disaster.

'The miners won't stand for Sven being killed,' Annie said. 'There's going to be trouble in Longyearbyen.'

On the Monday morning General Stolypin received a number of other disturbing local reports. Two border police had been insulted last night, a Russian-owned car had been set on fire, crude posters had appeared on walls saying

'Murderers Go Home'. He himself had noticed the scowls on miners' faces as he was driven the short distance from his commandeered house to the Governor's building.

Gazing out of the window over the melting ice in the fjord, mechanically drawing on one of his favourite cigarettes, Stolypin considered the implications of all this. The powers he had been given deliberately parallelled those of Sysselmann Prebensen. He was a judge as well as an administrator. He pencilled a few notes on a scratch pad, realized he was merely doodling, then came to a decision.

The incident had taken place in a Russian settlement. The sergeant was already under arrest. He would be charged with manslaughter and flown back to Murmansk for trial. The Russian court would not condone such a killing. The only concealment would be of his military rank, which was why there could be no question of a Norwegian trial. As for the team manager, he deserved to be shot for his stupidity. He would be on the next plane too.

Having made up his mind, Stolypin went along the corridor to see the Sysselmann, shouting for his interpreter as he went. He intended to negotiate this on a personal man-to-man basis, even though his speech would have to be translated.

'What can I do for you, General?' Prebensen asked coldly as Stolypin entered. He too had received a report on the incident, though it was overlaid by instructions from Oslo, just decoded, of far greater importance. It occurred to him that the Russians might have already intercepted his orders. Either way, today the situation in Longyearbyen was changed and he felt no hesitation in showing his anger.

'About this unfortunate affair at the football match.' Stolypin took a chair as if settling down with an old friend.

'Do be seated,' Prebensen said sardonically to the interpreter, a small man who habitually wore a badly cut and tightly buttoned blue suit as though his status depended upon it.

'The offender will be brought to trial,' Stolypin continued.

'In Norway, I hope.'

'The incident took place at our mining company's township. The case must be tried by a Russian judge. I assure you, Governor, the full rigour of our civil code will be applied.' Seeing little amelioration in Prebensen's expression, Stolypin put his main point as gently as he could. 'There was no excuse for the man's conduct,' he continued, 'even though the fight was started by your player.'

'Who cannot answer the charge because he is dead!'

'True,' Stolypin conceded reluctantly. 'However, there are many witnesses. What matters is for both sides to acknowledge their respective blame. Only in that way can we preserve the essential good relations between our two communities.' He sat quiet, letting the interpreter translate. He had been conciliatory to a quite unprecedented extent. Now it was the Norwegian's turn.

'You can do whatever you like for good relations,' Prebensen said. 'They are in your hands now.'

Stolypin looked at him questioningly.

'Beyond demanding that the murderer be brought for trial before a Norwegian court, I shall do nothing.'

'Please clarify what you are saying.'

Prebensen drew out a short text and handed it across the table. 'This is a copy of my government's formal demand that you, your staff and all Soviet personnel leave Longyearbyen, Svea, the Isfjord radio station and all other territory not leased to the Trust Arktikugol. If you do not comply by midnight tomorrow I am under orders to withdraw myself with my staff the next day. My government is no longer prepared to participate in this charade of joint administration.' Prebensen stood up. He had held himself in check for many weeks and felt deep satisfaction at being able to let his anger and scorn show at last. 'You can either leave Svalbard or face the international consequences. Good day, General.'

For a few seconds Stolypin was too amazed to reply. He felt the blood flush his face, then, almost to his own surprise, he relaxed and burst into laughter. The phrase about 'international consequences' was totally absurd.

'Well, well, well,' he said softly. 'Every dog has its day, eh, Prebensen?' He motioned to the interpreter. 'Come on, Boris Mikhailovitch, let's get back to serious business.' It was the most preposterous bluff he had heard in years.

'I've never been worked so hard in my life!' Julia was complaining volubly to Sergeant Webb. 'Anyone would think war was going to break out next week.'

'Things are a trifle hectic, miss.'

Webb had scarcely spoken before the buzzer rang and the voice of General Anderson echoed through the intercom. 'Have Sandra come in, would you?'

'Thank God it's your turn,' Julia said as Sandra rose from her desk. 'Everything's gone mad since that ministerial meeting.'

Sandra picked up her shorthand pad, checked her pencil was sharp and opened the General's door, stopping short just inside when she saw he was talking to Brigadier Curtis, whose boyish face was more than usually serious.

'... the Norwegians are preparing for a retaliatory attack in Finnmark,' Anderson was saying. 'The ACE Mobile Force ought to be at six hours' readiness to move in there.' He gestured to Sandra to close the door. 'Officially it'll be because of Soviet troop movements around Murmansk.'

'When does the crunch come, sir?'

'Roughly five days.' He smiled wryly. 'John King's telling me the minimum I need to know.' He turned his attention to Sandra. 'We've some complicated stuff here, Sandra, and we have to get it out tonight. Can you stay late?'

'Yes, of course.' Her date with Erich flickered through her thoughts. He could hardly object. 'It won't matter if I'm late.'

'Thank you. We've a crisis on our hands right now. Check it all through and come back in if you've any queries.'

She gathered up the sheaf of papers, noticing that the Brigadier's original drafts had been extensively altered in Anderson's neat but cramped handwriting. That was a pointer to the importance of whatever this was. She excused herself and went to her own desk to make sure she could follow all the annotations before going to the word-processor room.

Twenty minutes later the Brigadier emerged and paused to look over her shoulder. 'Any problems?' he asked.

She pointed out a few minor queries.

'Well, the sooner you're able to start, the better,' he said in a friendly yet urgent voice. He liked Sandra and he and his wife invited her to their house often, sensing her loneliness. She was a competent and willing secretary, if unimaginative, and he was sorry that their attempts to match her with bachelor officers had always failed. No sex appeal, that was her trouble. He patted her shoulder gently. 'I'll be in my office if you want me.'

By the time she had keyed the whole text into the word processor Sandra realized that General Anderson had not been exaggerating when he spoke of a crisis. The military dispositions that he was recommending to the various governments began with details of a total sea blockade of Svalbard, being instituted immediately by the Norwegian, British and US navies. It went on to ask for a host of measures from every European NATO country. Many seemed minor, like securing fuel stocks at dispersal airfields. However, the overall message was plain. The chairman of the Military Committee was preparing for an armed confrontation with the Warsaw Pact in the next week and he wanted the member nations to go, as it were, to action stations. In justifying this request for near mobilization, he quoted an unpublished part of the Defence Planning Committee's decisions. 'The moment has come to take justified and adequate defensive measures in the face of this attempt by the Soviet Union to destroy the solidarity of the Alliance and infringe its territorial integrity.'

She only stopped in her long task to leave a phone message at the Duc d'Arenberg to tell Erich she would be very late.

When she finally parked her car in the Grand Sablon and walked to the restaurant she felt too tired to eat. Erich was waiting at their usual corner table and she saw at once how impatient he was. For once she felt like quarrelling with him, particularly since she knew she had at last got the background story he was always demanding.

'If you're going to be so bad-tempered,' she said with a spirit that surprised herself, 'you can jolly well wait to hear my news.' Then, just to tantalize, she added, 'My respected employers really do seem to have got the bit between their teeth this time.'

'They have which thing in their mouths? Please explain.'

She couldn't help laughing, especially as she hadn't intended to bait him. Poor darling, she wondered if he would ever understand colloquial English. Then, because he became really angry, she relented and started sketching an outline in words of what she had typed. To her astonishment he stopped her.

'Later,' he said softly, looking suspiciously at the waiter. 'Tell me later.' He raised his voice. 'What marvellous news. Now, why not have an omelette? You must eat something.'

Back at her apartment, he quizzed her endlessly, always returning to the same point. 'But why are they doing all this? There must be a definite reason.'

'It's because of the DPC, I suppose, darling.' She was longing to go to bed and to sleep. 'I didn't really think about that.' Then she remembered overhearing General Anderson talking to the Brigadier.

'Whatever happens you mustn't quote this, but he did say the crunch was coming in about five days and the Russians might retaliate.'

'How?'

'You do promise not to put this in the paper? They'd know at once where it came from.'

'My darling.' He laid heavy emphasis on the endearment. 'I swear not one word of this will appear in the magazine.' It quite amused him to tell the exact truth and be able to laugh at her stupidity, particularly after she had made fun of him in the restaurant. 'On my honour.'

'He said they might retaliate by attacking Finnmark. That's the northern tip of Norway, isn't it?'

'Hmm.' Erich considered this, turning over its significance in his mind. 'That was all he said?'

'Yes. I don't see why the Russians should though, do you?'

'No.' Well, he thought, it's my job to report, not evaluate. But by definition to retaliate was to react to what the other side had done. Therefore NATO was taking some action.

Erich lay awake beside her that night, digesting the information and compelling himself to memorize it. He crept out of her flat early, found a call box and telephoned the number reserved for emergencies.

Bear Island edged up above the horizon on Monday afternoon, its southern end a precipitous cliff with a conspicuous tooth of rock projecting from the sea below and a plume of cloud drifting from its crest.

'Halfway house,' remarked Weston cheerfully to Mydland, studying the cloud-hung outline of the island through binoculars. 'Now for the fun, if Colonel Peterson's right.' He scanned farther round the view, balancing himself against the side of the wheelhouse. The *Northern Light* was corkscrewing in a choppy sea. 'Looks like a whaler over there. But what's happened to the Russian Navy?'

'Waiting for us nearer Svalbard, I suppose,' Mydland answered philosophically. 'We should start fishing then.'

'Don't say that too loud. All the boys need now is to think there'll be half a ton of cod joining them down there.'

Weston had deliberately allowed a day and a half to cover the 240 nautical miles from the rendezvous with the destroyer to Bear Island. Even in that short period, the men had begun to hate the confinement of the hold. From Bear Island on, Weston planned the voyage to be even slower.

'You're right, Paul,' he acknowledged. 'When we reach the fishing grounds we'll bloody well have to put up a pretence.'

'We must call the Bear Island radio too. Ask for a weather report. That would be normal.'

They watched the murky silhouette of the island slowly harden into firm lines, revealing patches of snow down to the water's edge. Rounding the coast, the radio masts came into sight, thin spindles sticking up near a group of buildings on the flat northern part. Uninhabited except for the communications engineers, Bear Island was a key listening post.

'Can't understand the Russkies not taking over that lot,' said Weston. 'Seems an obvious thing to do.'

'That may be the next step.' Mydland had wondered the same thing. 'Perhaps they haven't done it because there are no Russians normally there, although they claim the 1945 defence agreement includes Bear Island. Let us make contact now.' He picked up the microphone of the ship's radio and conducted a short conversation in Norwegian, his expression becoming more and more concerned.

The wheelhouse door opened and Clifford came in, a woollen cap pulled down over his close-cropped head. He wore a blue seaman's jacket over his uniform.

'Heard you transmitting,' he said, when Mydland had finished. 'What's new?'

'A lot of Soviet naval activity north of here. Merchant shipping is advised to keep out of the area. Weather is good, however. Ice breaking up off Svalbard. Wind force three, sea state two. Visibility good.'

'Hell! Means we have to wait around some?'

'The weather changes very quickly in Svalbard. We have three days in hand.'

'Two days,' Weston corrected him. 'We still have a

couple of hundred miles of sea, remember.' He checked his watch. 'Time for lunch. Let's hope Sergeant Millar's turning up something good.'

He left the wheelhouse and went back to the cabin where the four officers and the Norwegian engineer, Paulsen, usually ate. Millar was busy in the tiny galley, wearing a thick hand-knitted sweater and stained trousers. Like Weston, he spoke some Norwegian and would, if the boat were stopped, play the part of a seaman, along with Paulsen.

'How goes it?' Weston asked.

'Hope to Christ they like this a bit better.' Millar grimaced. He had never anticipated that food would become a source of friction. The *Northern Light* had been provisioned in Tromso, since the presence of American products in the galley would immediately strike a wrong note to a visitor. Not that the ship could withstand a detailed inspection, but they might dupe a Soviet officer who only came on board cursorily. However, the unfamiliar rations, added to the cramped conditions in the hold, were already causing complaints.

'They'll have to like it,' Weston said.

'You're damn right they will.' Clifford had come in unobserved. 'It's not that they're picky,' he went on, taking a slice of the meat in his fingers and tasting it. 'Just they feel they could be having American chow.' He licked his fingers. 'This is great. Hell, Major, when they're training in Florida they get to eating snakes and lizards.'

'Better find a recipe for polar bear this time, sir,' said Millar good-humouredly. 'Personally I'm partial to a reindeer steak.'

Clifford laughed. 'Let's serve this up. They're going to be on "lurps" soon enough.'

The two officers sat down on the oilcloth-covered bench by the cabin's table and began eating, while Millar called the men to come up one at a time and get their food. Part of the modification Weston had carried out to the ship was to cut a way through from the hold to the engine room, whence there was access up by iron rungs to the cabin superstructure. For safety reasons the narrow aperture had a watertight door fitted.

The two majors discussed plans as they ate. The landing on Svalbard, using a normal commercially made inflatable rubber boat which was kept packed up in a large bag, would have to wait for cloudy weather, preferably fog. Sergeant Millar and one American would form the beach party, going in first, while Weston remained on the ship.

'Should be a piece of cake,' Weston said. 'Just so long as we've got poor visibility and no Russkie patrol craft are around.' He had rapidly warmed to Clifford, a man superficially as different from him as could be imagined. While Weston, puffing his pipe in his oil-stained white sweater, managed to appear so relaxed and confident as to be unnecessarily casual, Clifford spoke incisively and moved with the briskness of an athlete.

For his part Clifford now appreciated that Weston's flippancy was merely a front. Underneath he was both competent and calculating. Clifford reckoned he could be relied on to get them there and to be available with the trawler if the basic withdrawal plan went wrong.

They were chewing over this crucial aspect of the raid when Mydland's head appeared in the door. 'Helicopter approaching,' he announced. 'Not one of ours, I think.'

They stopped talking instantly and went out on deck. Mydland was right. A helicopter was coming in towards them low from the east, though still inaudible over the deep thumping of the trawler's diesels and the swish of water along the hull.

'Better get below, Craig,' Weston advised. 'There's no way you'll pass as a Norwegian. Even at fifty yards.'

'Sure.' Clifford disappeared down. Weston closed the hatch and went into the wheelhouse, then re-emerged as the helicopter came overhead, gazing at it as a sailor would. The paint scheme and the markings established that it belonged to Soviet Naval Aviation. He watched it circle the *Northern Light* several times. Then its pilot impudently hovered overhead, and the downwash from the rotors sent papers inside the wheelhouse whirling off the chart table. Weston slammed the door shut and gestured angrily. Quite clearly he saw the pilot raise an arm and flip his hand downwards, contemptuously dismissing the complaint. The machine lurched into forward flight and left them.

'Bloody nerve,' Weston muttered. He went inside. 'We'll be getting a visit soon,' he told Mydland, 'or I'm a Dutchman. That was a Kamov, probably from a helicopter cruiser like the *Moskva*. We'd better stand by for boarders.'

An hour later the blip of a fast-moving ship appeared on the radar screen and soon after that Weston identified it through his binoculars: the lean, aggressive shape of a destroyer with an easily recognizable funnel and superstructure, crowned by a mass of aerials.

'Kresta class,' he said nonchalantly to Clifford, who was back up in the wheelhouse. 'And doing at least twenty knots. Must be in a hurry to make our acquaintance.'

'Just blow the whistle when you need us.'

'Don't worry, we'll fool them.' Weston pulled an old blue peaked cap on to his head, relit his pipe and took a turn around the deck to check if everything looked normal. It did.

When the destroyer drew close, he became apprehensive, remembering how another Russian ship had once taken such risks whilst shadowing a NATO naval exercise that it had collided with a British aircraft carrier, hugely bigger than itself. If this captain had orders to harass Norwegian merchant shipping he would cheerfully ram a trawler, knowing that even if it sank his government would claim it was the trawler's fault.

Weston brought the *Northern Light*'s engines back to slow ahead, losing way as she rocked in the swell. The destroyer was fifty yards off now, and he could see the officers on the bridge. A coloured flag was hauled up by a seaman. Before he could even distinguish its design, the destroyer swung across his bows. It was moving faster than the trawler. He spun the wheel hard to port, but the response was too sluggish. The grey bows of the destroyer reared up above him and with a shuddering crunch the two ships collided, the impact throwing the *Northern Light* sideways with a fearful sound of wood splintering. A moment later it settled back and ground against the steel plates of the destroyer, whose engines were now churning the water white as it went full astern.

An officer in greatcoat and high peaked hat appeared on the outer part of the bridge, holding a megaphone.

'What are you fools doing?' he shouted in Norwegian, pointing to the flag fluttering in the wind. 'Can you not recognize international signals? You were ordered to heave to for inspection.'

'You have no right to stop us,' shouted Mydland, now out on deck too. 'You could have sunk us.' He shook his fists at the Russian in unfeigned anger.

The officer took no notice, but scanned the *Northern Light* from the blue and red Norwegian flag hanging limp at her stern to the nets and ropes piled forward of the hold in the bow.

'What are you doing here?' he demanded through the megaphone.

'Going to fish, damn you.'

Again the officer did not reply. Another joined him, raised a camera and took a succession of photographs.

'Complain through your embassy if you want,' the officer called down. 'You should be more careful in these waters.'

He vanished inside the bridge and they heard the rumble as the engines gained power. The warship swung away and left them.

Weston leaned cautiously over the bent and broken rail and tried to survey the damage. The timbers where the deck joined the hull were cracked, though not completely stove in. Below the water line they might well be leaking. He hurried back through the cabin, down the steel rungs to the engine room and through the door to the hold.

Inside was confusion. Clothing and kit were strewn on the floor, thrown off shelves and bunks by the impact, though there was no sign of water coming in. The men stood pressed to the sides of the hold's narrow confines, silently watching the corpsman Johnson, who was kneeling by one of the lower bunks. On the bed lay Clifford, partially propped up on his elbows, his face in shadow. The area was lit by only two electric lights.

'What's wrong?' Weston asked.

'The Major fell, sir.' Johnson looked round briefly, then resumed his examination of Clifford's left leg, from which he had removed the boot and sock.

Weston leaned over. He could see the ankle was distorted and the flesh seemed swollen.

'Fractured,' Johnson said, feeling round the bones with his fingers. Clifford winced but did not speak. 'Guess the best thing I can do is bandage it.' He reached for a small rucksack, pulled out a metal box with a red cross painted on its top and foraged among the carefully packed tubes and containers inside. 'Should be in plaster, but that's a doctor's job. I don't know how to set breaks. You need a medevac, sir.'

'Just have to wait,' Clifford commented tautly, pain edging into his voice. He glanced at Weston. 'What the hell happened?'

'They rammed us. Deliberately.'

'Bastards!' said Howard Smith loudly from the corner where he was perched on a top bunk. 'But it's going to take more than a destroyer to stop this raid.' He spoke loudly, as though asserting the leadership he would now assume.

'Damn right,' Clifford said. 'Listen, Corporal, you just fix this little problem so as I can walk, will ya?'

Johnson shook his head. 'Can't do that, sir. No way you can march on that ankle. You touch it, you can tell where the bone's gone.'

Clifford hoisted himself as upright as the bunk above would permit, twisting round and forward until he could reach his foot. Then he grunted. 'Guess you'll be taking the boys into the promised land, Howard,' he conceded. He consulted Weston again. 'Might as well face up to all the unpleasant facts. Could they have rumbled us? Or does sinking trawlers come naturally to them?'

'My guess is they're reminding every Norwegian ship around that they own the Barents Sea. Or think they do. Simple bully-boy tactics.'

'I sure hope you're right.'

'The Soviets must have surveillance on everything that moves up here,' Smith cut in. 'From satellites, aircraft, their own navy. Maybe we're going a shade too direct toward Svalbard.'

'I agree,' Weston said. 'We're going to start fishing for a few hours shortly. Catch some cod for supper.'

He raised a hand in casual salute to Clifford and left the hold, feeling in a far from joking mood. He was worried about what further attentions the *Northern Light* might receive from the Soviet Navy, especially if the US Fleet implemented its threatened blockade of Svalbard.

After supervising the transfer of the raiding team to the *Northern Light* on Saturday, Peterson had been lifted by helicopter to the Norwegian airfield at Bodo and driven to the Reitan headquarters. There he had been greeted by the Norwegian major general who, he now discovered, had been directing Paul Mydland's activities since the Russian takeover.

The revelation, made over a quiet meal in a private dining room, caused him considerable surprise.

'Hell!' he exclaimed. 'So he's already seen a piece of the action?'

'You could say so,' the general answered drily, then added, 'We didn't wish to advertise the fact even among Allies. It is a considerable risk for him. He has already been interrogated by the Russians once. However, he does know the way.'

When Peterson pressed for more information, the general avoided answering in precise terms. 'If I have one criticism of your armed forces,' he observed eventually, 'it is that too many people have the right to ask what they are doing, from your Congressional Committee onwards. Here in Norway we also act within very difficult political constraints, but we are allowed to keep our secrets to ourselves. As for NATO – my God! I served there once. The intrigue and jealousy are unbelievable.'

'You think we're on a hiding to nothing with this operation, sir?' Peterson brought his fear into the open.

'For myself – ' he was weighing his words carefully ' – I might have searched for other solutions. Perhaps depended more on the blockade.' He fiddled with a small glass of the faintly brown-coloured acquavit that accompanied their meal. 'You know, it was Lenin himself who acknowledged our sovereignty in Svalbard. There is still leverage in that fact.' He raised his glass. 'Well, the time is past for debate. We can only wish your team luck. They will need it and more. Let us say "skol" to them, shall we?' He clicked his glass against Peterson's and downed the liquor in a single gulp. 'And if you want any material help in the next few days, please ask for it.'

'As of today I'm pretty much a spectator,' Peterson said, feeling the acquavit scorch his throat yet deriving no reassurance from it.

The general wiped his mouth neatly with a linen napkin. 'Would you care for a short drive? There is a hill close by where one has a magnificent view of the midnight sun.'

Later, as they stood watching the orange-gold ball hovering on the northern horizon above the endless sea, Peterson felt even more strongly how isolated he now was from the raid he had planned. It was almost like parting from a loved one: afterwards you think of the many things you failed to say. They were going into action, while he returned to the safety of an operations room. He had never thought of himself as one of the pogues in the rear.

'The planners have to stay behind,' said the general, appreciating his silence. 'They must put their judgment on the line.' He laughed quietly. 'Which is one reason why promotion to the higher ranks is always so uncertain.'

Two days later Peterson remembered this conversation more acutely than he anticipated at the time. From Reitan the Commander North Norway had instant communication not merely to Oslo, but also to SHAPE and to SACLANT. Communications have always been the lifeblood of warfare: the disastrous if heroic Charge of the Light Brigade in the Crimea resulted from a misunderstood message. Ever since its creation NATO had steadily improved both the Alliance's voice and data links and their security. In consequence the duty officers at Norfolk and Reitan received the coded signal from the *Northern Light* almost simultaneously.

'Here, sir.' The Norwegian handed Peterson the decoded transcript on a message form. 'Looks as if they had a near shave.'

Peterson read the terse signal twice, then referred to the *Northern Light*'s position on the illuminated display which ran along one wall. The trawler was some 60 nautical miles north of Bear Island. Her landfall was due late Wednesday night or in the early hours of Thursday, at Weston's and Clifford's discretion. She had less than 170 miles to sail in over two days. She could afford to dawdle, to cast her nets, to feint in wrong directions. Ploughing along at 10 knots, she had roughly thirty-eight hours extra in hand. A lot could be done in thirty-eight hours.

The idea came to him instantly, born of his lack of complete faith in Howard Smith, of the time in hand, of the Norwegian general's remarks about the team needing luck and then some. He told himself not to be a fool, every member of the team could double for another. Would he himself bring them any advantage? For a few minutes he cross-examined himself as to his motives. Then he drafted an urgent signal to Admiral King at Norfolk.

Forty minutes later King was on the secure line in person.

'Are you crazy, Peterson?' he demanded. 'How are you going to get there?'

'Free fall, sir.'

'And have every damn Russian radar spot the plane?'

'They won't notice anything different about this one. If the Norwegians play ball.'

'Maybe. I'll think about it. Come back to me with a plan fast. And Peterson?'

'Sir.'

'Didn't you select the men yourself?'

'I wasn't given that option, sir. And the team needs a more senior man now Clifford's out of action.'

When the Admiral had rung off, Peterson set about organizing himself. The results were more encouraging than he could have hoped. A Scandinavian Airlines DC-9 was being sent into Longyearbyen empty on Wednesday morning to collect the Sysselmann and his staff, who were being withdrawn from Svalbard. His next call was to the headquarters of the Special Forces, which had developed the technique of parachuting from the rear entry stairs of an airliner at altitude after a hijacker had done just that in America, apparently surviving complete with his ransom money. The point of this concept was simple. Assuming you could obtain the use of a civil aircraft proceeding over your target territory on a normal flight plan, the plane's presence would arouse no suspicion, and you could parachute from it with little chance of being detected. All you had to do was free-fall most of the way down, only opening the canopy a couple of thousand feet above the ground. You needed oxygen, of course, and parachuting skill, and a crew who would keep their mouths shut.

By the evening Admiral King had agreed in principle. He too was concerned at the raid going in without its commanding officer, whatever the Delta Force's normal operating principles were. Peterson would join the airliner in Oslo, ostensibly as a crew member.

That evening Peterson put in a call to Nancy in London. He hesitated some time before doing so, fearful that she would pick up some trace of tension or excitement in his voice and be alarmed. Then it occurred to him that, if everything went wrong, this might be the last time he spoke to her, so he took the chance.

Sure enough, his opening greeting of 'How are things?' sounded false even to him.

'Where are you, Tom?'

'Still in Norway. Don't worry, everything's fine.'

'How can I not worry?' She was finding her enforced idleness almost more than she could bear. 'Will there be a war? The newspapers say we and the Russians are eyeball to eyeball.'

'Makes a good story, but don't believe everything you read, sweetheart. Sure there's a crisis, we just have to keep our nerve, that's all. Now, let's talk about something that matters. Are you OK?'

'Oh yes.' She couldn't make the answer enthusiastic. 'The morning sickness has stopped. I feel fine. I just wish we were both back in Virginia Beach.'

'Don't worry, sweetheart, we will be. I should have finished here in about a week.'

His firm tone reassured her. She felt a surge of happiness. 'Well, let me tell you something, Tom Peterson, you'd better be because that Admiral of yours is going to have hell to pay if you aren't.'

'I will be.' One of those sick jokes the grunts used to crack in the Nam echoed in his mind. 'You know, Lieutenant,' a man would say before an operation, 'if I get wounded on this one it'll be bad enough, Ma crying and the whole bad scene. But if I get killed I'll really be pissed.'

'I love you and I'll be back just as soon as I can,' he promised. But when he went to bed he couldn't sleep, his closed eyes seeing images of the desolate snowscape coming up to meet him as he squinted down from a parachute harness, and the whine of the jet's engines fading to silence.

7

Julia noticed the ring at once: three sapphires and two diamonds forming a chequered band, glittering on the third finger of Sandra's right hand.

'Very pretty, darling,' she commented, not even trying to keep the envy out of her voice. 'Very pretty indeed.'

Sandra blushed. 'You really think so? Erich produced it from nowhere this weekend. Well, not quite that, but it was a surprise.'

'So much the better. I'm all for surprises, especially diamond ones. Let me see it properly then!' She examined it closely on Sandra's outstretched fingers, thinking that unless the stones were paste young Erich must have more money than anyone supposed. She felt quite jealous enough to be genially nasty. 'Lucky you. But why aren't you wearing it on the other hand?'

'We're not announcing anything yet.' The colour stayed high in Sandra's normally pale cheeks. She gazed at the ring herself, wishing it could answer some of her doubts, though she could understand Erich not wanting a formal engagement until she had met his mother. 'It is lovely, isn't it?' she said almost wistfully. 'We saw one I liked in Deauville and this is practically identical.'

'Ah! I guessed something special happened that weekend.' Julia winked one heavily painted eyelid knowingly.

They were interrupted by Sergeant Webb with an urgent signal. 'Sorry to curtail your exchange of views, ladies, but it ain't Monday morning for nothing.'

At lunchtime Julia joined Sandra in the canteen. She was deeply curious about Erich, whom she had marked down as a first-class stud but selfish, ungenerous, inquisitive – too inquisitive. Why in heaven's name should he want to tie himself to a girl who hadn't even learned to hold herself properly and had no discernible sex appeal? Did he think Sandra was an heiress?

When they had solemnly devoured their yoghurt and cheese salads, each invoking their diets, Julia was little the wiser. What she had discovered was mainly from things left unsaid. Sandra seemed apprehensive as well as happy, speaking as though her relationship with Erich was too precious to discuss. All she revealed was that when he had his hoped-for new job in the autumn they would move to Berlin.

The mention of the divided city, so long a focal point of the undeclared war between the Soviet empire and the free world, triggered memories for Julia. She had an immediate vision of Ursula, the West German secretary who had worked in the same room as Sandra now did. Ursula had been inspired by such typically Germanic devotion to duty that she never minded working late and so interested in her job that she made any excuse to visit the Military Committee's situation room. She had even laboriously translated the words on a Roman carving in the passage outside. When she rang one Monday morning to say she was a little ill, no one questioned it. Ursula wasn't the sort to go sick deliberately. Two days later they wondered if she was all right and sent someone to her flat. She wasn't there. Too late they discovered that she was perfectly well, but in East Berlin. She had been an agent of the Stasi all along, and fled when the Stasi thought her cover had been blown.

From which part of Berlin did Erich come, Julia suddenly wondered, East or West? Why did he keep Sandra so in the dark about his past? Could Sandra, without knowing it, be

Ursula's replacement, manipulated by a demanding lover? The suspicion was inescapable, at least to a woman of Julia's cynical nature.

Late in the afternoon, when Brigadier Curtis was at a meeting, Julia slipped into his office and dialled London on the phone. She spoke to a man called Phillpotts. He remembered Sandra instantly from his last NATO visit.

'She's become engaged to a German journalist who works for a magazine in Munich called *Aktuelle Nachrichten*.'

'Has she, by Jove?' Phillpotts's manner of speech was markedly old-fashioned. 'A Kraut, eh? Ought to have told us, silly girl.' Security procedures were a lot more stringent than they used to be.

'I think there's something wrong about him.' Julia gave a brief description of Erich's age and appearance.

At the other end of the line Phillpotts noted down the facts. '*Aktuelle Nachrichten*, you said?' That name rang a bell and he needed no reminding that the Military Committee was at the epicentre of the present crisis. 'I'll come across as soon as I can, tomorrow or Wednesday. Thank you, my dear, thank you very much.'

Julia put the phone down hoping she had done the right thing. He was a funny old bird, Phillpotts, looked more like a retired cavalry colonel than a security officer but was shrewder than he seemed. Each country was responsible for the security clearances of its nationals on the NATO staff. He did the checks on the British secretaries. They regarded him as some kind of benevolent uncle.

Two hundred miles away in his London office Phillpotts began making inquiries with remarkable speed. There were two sides to Julia, as he knew well. If he had not come to respect her basic integrity, she would have been sacked long ago. 'Bet five pounds to a horseshoe she's been to bed with the bloody man herself,' he muttered as he initiated a trace request on Erich Braun. The West German Security Service might know something both about him and the magazine.

*

The DC-9 stood parked away from the main terminal at Oslo's Fornebu airport, a sleek white ghost with the national flags of Denmark, Norway and Sweden emblazoned on its tail. The authorities had managed to keep the airliner comparatively speaking in the shadows; not that the requirements of security left any part of the tarmac unlit.

As Peterson walked out with the captain he felt exposed and uncomfortable. His borrowed peaked hat and blue pilot's uniform fitted tolerably well. Scandinavian Airlines aircrew mostly wore their hair short as he did, and he had the right kind of lean, alert features for the part. Even the pilot's badge on his chest glittered dully at the extremity of his vision just the way the silver airborne wings did above the medal ribbons on his Marine Corps tunic. Despite all this, he felt awkward and apprehensive of being asked some technical question by the ground crew.

'We should check those rear steps once again,' remarked the captain, leading him towards the tail, where the rear belly of the plane opened downwards and backwards to provide a narrow stairway for passengers. The DC-9's designers had of course never intended the steps to be extended in flight. Peterson climbed halfway up the steps, examining the hydraulically operated steel pistons on each side which held the stairway in position.

'Just walk down and wave goodbye, I guess,' he said with a light-heartedness he did not feel. 'Not too different from the tail ramp of a C-130.' He had done plenty of free-fall jumps from the lumbering military transports.

'You think so?' The captain was surprised. He had one of those weathered, experienced expressions so beloved of advertising photographers and the lines around his eyes crinkled closer as he questioned this assumption. 'Very different, I should say. Faster, for one thing. I will need 115 knots for safety.'

Peterson imagined the slipstream tearing round him. He wasn't sure he could make a clean exit laden with equipment, trying to descend a narrow aluminium stairway in a wind three times gale force. 'Can't you make it any less?'

'With some flap and the gear down, maybe,' the pilot conceded. 'In landing configuration perhaps between 105 and 110 knots.'

'You'll be low and there'll be no pressurization problems.'

'I suppose we will manage.' The captain dismissed the question and turned back to Peterson. 'Let's go up front and have a further look at the map. Dropping you accurately may not be so easy.'

'Just so long as we don't come down in the sea.'

'If there is any chance of that, we must abort the mission. You would not survive the cold.'

When they were on the flight deck he spread out a map of Svalbard so far as the confined space allowed. Svalbard's main island, Spitsbergen, was tooth-shaped, with the tip to the south. Cutting into its west coast were a number of fjords. The biggest of these were the Bellsund and the Isfjord, separated by a 25-mile-wide chunk of mountainous land. Both the Russian settlement of Barentsburg and Longyearbyen itself lay on the northern side of this chunk, while the map showed two radio stations on the coast. The captain jabbed his finger at them.

'The Bellsund radio and the Isfjord radio both give us good fixes. You want to be how far from the Bellsund station?'

'Five miles roughly. Eight kilometres.' Peterson indicated a wide bay named Marvagen. 'Inshore from there.'

'Which means you have a good space between the mountain and the sea.' The captain studied the area, fingering a pocket calculator. 'At 110 knots we should put you out two minutes and twenty seconds after passing Bellsund. Our normal altitude would be around 5000 feet at this point.'

'That's fine. There's Soviet radar will be tracking us, but I did some checks on the intervisibility.' Peterson produced a folded sheet of graph paper on which he had marked mountain heights and distances. 'If you can drop me within two miles either way of the place I want, then we're in business. They have radar 1500 feet up above Longyearbyen but at that point there'll be a line of higher peaks between us and it.' He consulted the graph. 'I guess they'll lose us at any height below 6500 feet. Same goes for the radar on the shore near Barentsburg. We'll be screened from it, too.'

'Can they pick up a parachutist in free fall, anyway?'

'Only if they're extremely alert. The satellite cameras can't, if it's cloudy.'

The captain considered this. 'And what if you cannot see the ground?'

'All the better for concealment. Can't say I've ever tried free-falling through cloud, in fact it's illegal in the States, but I guess there's a first time for everything.' He was determined to be optimistic. This was his scheme and he was damn well going to make it work.

'Rather you than me,' said the captain, admiration in his voice. 'Well, we shall take off in two hours' time. Land at Tromso to refuel 0400 and should be over Svalbard about 0640. It will be daylight, of course. We have to hope there are no other aircraft in the vicinity. If there are, cloud could be useful.'

The flight was uneventful. Peterson dozed, soporific because it was midnight, letting both body and senses droop in the way airborne soldiers always do on the long flight to a jump, yet unable to stop memories racking his mind. He imagined Nancy in the tiny London apartment, wondered if she was asleep, how much of his mission she divined. She was there, alive, untouched in a world he had temporarily left behind. It occurred to him to write her a letter, just in case it was his last, and he shook himself fully awake to get paper from one of the cabin crew. After a few scribbles he tore up the result and started again, ending with a simple, 'Be back in a few days. Take care. I love you. Tom.' He sealed it in an envelope and gave it to the captain to be posted, wondering if it would be censored. The crew, observing him with open curiosity, had a surprise coming on their return. He had been assured they had not only been sworn to secrecy, they would be kept in comfortable isolation until the operation was over. There had been too many past cases of espionage in Norway for the government to take any chances now it had agreed to action.

Once they had taken off from Tromso, Peterson zipped himself into a heavily lined white jump suit, warm enough to be a substitute for a sleeping bag on the ground. The stewards watched, fascinated, as he checked his equipment. He had his maps coated with a clear laminate and folded them into a long pocket on his trouser leg. He carried his favourite weapon, a .357 magnum revolver, ammunition and a small strobe light, all attached to his war belt. The parachute was buckled on his back. It was fitted with an automatic barometric opening device which would deploy the canopy at 2000 feet if he had not already done so. Attached to the reserve parachute harness on his chest was an altimeter.

Fifteen minutes out from Svalbard he strapped to his right leg a long container which eventually would hang below him on a length of nylon rope. Inside it were rations, spare clothing, an M-16 carbine, folding skis and a small radio. Finally he clamped a helmet on his head and swayed awkwardly towards the rear, like a figure out of space fiction, balancing himself in the aisle between the seats.

The sergeant who would despatch him, also in jump suit and parachute in case of accident, waited at the telephone normally used by the cabin crew.

'Longyearbyen report cloudbase 200 feet. The captain asks if that's OK.'

'That's fine,' Peterson answered. He wasn't aborting now. They sensed the change in the plane's altitude as the captain reduced speed. Their ears popped as the cabin was depressurized. Through the windows they could see only cloud. There was a grinding whine as the pilot lowered the flaps and a jolt as the undercarriage came down.

At the end of the plane, between the lavatories, was a

door. The sergeant opened it inwards, revealing a conical cavity. He reached along the right-hand side, slipped open a guard and pulled a lever. The floor in front of him began to drop, icy air came blowing in with a great rushing noise, and the whine of the engines grew deafening. Peterson pulled the goggles down over his eyes, unable to see much beyond the sergeant. As the hidden steps extended the whole aircraft began to vibrate. Loose pieces of paper fluttered around the cabin and the stewardesses huddled in their seats at the front, tightening their seat belts apprehensively.

The sergeant stood with the telephone at his ear, then lifted his right hand and raised two fingers, the signal for 'Action stations'.

Peterson shuffled forward and poised himself in front of the gaping stairway, the sergeant squeezing to one side to let him past.

'Go!' The shout was inaudible, but Peterson saw the single raised finger and launched himself forward.

He tried to run down the steps, was caught by the blast of air and tumbled into the void, turning over as he fell. For a second he was totally disoriented, blanketed in cloud, then instinctively he shifted his limbs into the correct position, legs out, arms half folded as though crouching on his palms. The rush of wind increased. He achieved stability, head slightly down, able to control his 'flight' by small movements of his forearms. He could see the altimeter clearly: 3200, 3000, 2800, it would soon be unwinding at a rate of 10,000 feet a minute.

Four seconds later he pulled the chrome handle, the harness jerked his shoulders, he swung forward and looked up. Above him the dark fabric blossomed like a rectangle of eiderdown in the fog. This sophisticated parachute was designed more as an inflated wing than a canopy. The altimeter read 1900. Now he was falling at only 16 feet per second. He could hear the drone of the departing DC-9, but he saw nothing. He snapped free the hooks holding the container. It fell away and hung below him.

That, he realized with a flicker of fear, was the last positive action he could take, short of jettisoning the container altogether if he did find himself over water. He concentrated on controlling the parachute, trying to descend vertically rather than forwards. A controllable chute like this could travel fast. In clear weather he would have steered himself towards a chosen landing zone. But the cloud was as thick as fog. His only option was to trust the DC-9 captain's navigation. Drops of moisture began to freeze on his face. He peered down, checking the altimeter frequently.

At 250 feet the white obscurity shredded into tendrils of cloud, which quickly parted to reveal a deserted expanse of snow below. He had just time to glance all round and then the ground was coming up at him. He pulled the toggles, turned the canopy into wind, bent his knees and landed softly, not even rolling over.

The parachute, caught by the gusting breeze, tugged at him. He hauled in the lines, collapsing it, then bundled it up, pulled out the revolver and took stock. There was no one in sight. He took careful bearings with his compass. To the east the snow slope vanished upwards into cloud, perhaps two miles off. The sea must be a similar distance in the other direction. That was the way to his RV at beach Yellow, where Mydland would be waiting.

He hauled the container to him by its rope, unpacked it and prepared his kit for the short march. Then he buried the parachute in the nearest snowdrift together with the container, fitted his skis and set off, his white-covered pack on his back, the M-16 slung over his shoulder. So far, so good, he thought. The team must have heard the aircraft, they'd know he was on his way.

Nancy Peterson kneaded her eyes gently with her forefingers, as if she could rub away the gritty weariness of a restless night. The sunlight glared through the thin curtains. She shook herself, then glanced sideways at the travelling clock on the bedside table. Six-fifty. She felt terrible. How could a night pass so slowly, in such loneliness, tormented by such fears? She struggled out of bed, slipped on a bathrobe, and walked barefoot to the tiny kitchen to fetch juice from the icebox. A transistor radio stood on the worktop. She switched it on, pop music blared tinnily and she hastily reduced the volume. The song ended and a slick, bright, mid-Atlantic voice cut in, determinedly enlivening the breakfasts of eight million Britons.

'. . . and after the weather and news we'll be back at ten past seven with music from the Jackie Elton Four and . . .'

The brief weather report forecast Britain's predictable 'showers and sunny intervals' while Nancy made herself coffee. Then a third voice came on, fractionally more serious in tone, but still insistently cheerful.

Now the news at seven o'clock on Wednesday 12 June. As the Svalbard crisis deepens, ships of the US Atlantic Fleet are reported to be in close confrontation with the Russian Navy in the Barents Sea, following yesterday's statement by the American President that the establishment of early warning radar by the Soviets in Svalbard was totally unacceptable to the United States. Informed sources in Washington believe this could be the prelude to a full-scale naval blockade of the archipelago, where the few ports will soon be clear of ice.

In Moscow the newspaper *Pravda* today publishes its strongest condemnation of United States actions yet. Clearly expressing the policy of the Politburo, it attacks America's blatantly aggressive posture and warns that the President should not expect a repeat of the 1962 Cuban missile crisis, when the Soviet Union was forced to turn back ships carrying missiles to Cuba. This time the Soviet Navy is operating in home waters, and the United States should withdraw from the Barents Sea before it is too late. If not, the Soviet Union will be compelled to take retaliatory steps.

In London another massive anti-war demonstration is planned for this afternoon as Parliament debates the Government's recall of selected reservists to the Army, under a proclamation signed yesterday by the Queen.

Throughout Western Europe, demonstrations, strikes and

protests continue to dominate the headlines. In West Berlin, police with water cannon dispersed a crowd of fifteen thousand –

Nancy flipped the set off and dejectedly perched on a stool. Svalbard was where Tom had gone, she was sure of it. More was going on than either side was admitting, the two powers couldn't just stay deadlocked. They had to act or back down, and Tom must be involved. Look at the man's past, darn it. The Marines, the CIA, the Delta Force. She shivered. Where was she now, the woman who wasn't going to be the typical Service wife waiting home? Waiting in a rented apartment in a foreign city! Oh God, she thought, what can I do? Slowly, as her coffee grew cold, she understood. There was nothing she could do, no one she could call, at least no one who would tell her anything, and if she went home, back to Virginia Beach, it would be almost worse. There people would understand even less what was happening. Plus Tom was expecting her to be here when he came back. If he came back.

She stifled tears, blew her nose on a tissue and decided to stay put. She would listen to the strange accents on the radio and read the unfamiliar British papers and pray. She would try to think of a name for the baby, too, a name Tom would like. The baby might be all she had left of him. It was not an idea she wanted to contemplate. She forced herself to shower and dress and go out to look for a copy of the *Herald Tribune*.

No land was visible from the wheelhouse of the *Northern Light*, only mist and choppy grey water. However, Weston could trace the shoreline on the radar screen. The beam of yellow light, methodically rotating, illuminated the pattern of the bay near which Peterson ought to have landed four hours ago. It showed the headland of Kapp Martin where the Bellsund radio was situated and revealed the savagely steep mountains beyond. On the ship's radio, the kind most fishing vessels carried, he could hear the Bellsund's automatic Morse signal, the letters MT, endlessly repeated. At this distance he did not expect problems with short-range VHF reception. He knew precisely where he was, namely six miles west of Kapp Martin, maintaining a sluggish three knots and keeping a keen lookout for the floating chunks of ice which appeared to be drifting down from the north again. But where the hell was Peterson? He lifted his head from the radar's protective shield and spoke to Clifford, who was sitting on a small bench, keeping his injured leg off the floor.

'We have to assume our last message was not received. We're completely out of contact with Oslo. Could they have cancelled the drop?'

'I doubt it. Colonel Peterson isn't the quitting kind. Besides, we heard the plane go over.'

Weston considered the point. His long-range coded communications had become progressively more subject to atmospheric interference as the weather deteriorated. He had received the instructions about the drop, and his reply had been acknowledged. There were two possible landing places for the team: one on the Marvagen bay, the other round in the Bellsund itself, much closer to the cached stores and the cover which the mountains would afford. Peterson had designated code names for them, Yellow and Red.

The raiding party, using its 'burst' radio, would notify Bodo of a successful RV and Bodo in turn would authorize Weston to clear the area. Given the terrain, Peterson had no option. He had to jump near Yellow, but if the weather had suddenly cleared, the team would have been appallingly exposed to observation both from sea and air while waiting on the bay. When the mist came down yesterday evening the opportunity to land undetected was too good to miss. Weston had edged the trawler into the mouth of the Bellsund and the team had gone ashore in the Zodiac inflatable boat at Red, a spit of land roughly eight miles from Yellow. They had gone in two groups, Sergeant Millar first with one other as a beach party, the second with Clifford, who, despite his ankle, then brought the Zodiac back and reported complete success. The men scrambled onto a foreshore littered with ice which miraculously had not damaged the boat. Weston had then transmitted the briefest possible signal to Bodo: 'Red now, repeat, Red now.' There had been no answer.

'They can't have got it,' Weston said. 'We'll have to break radio silence. If the cloud lifts he'll be a sitting duck.'

Talking to Peterson on the VHF was a very different matter to the highly secure, fraction of a second, communication with Bodo. If the Russians had any kind of direction-finding system operating in the area, VHF traffic would enable them to locate Peterson and the *Northern Light*. True, they would have to be monitoring a large number of frequencies, but they had the equipment. Somewhere up above the Arctic, day in, day out, through periods of peaceful coexistence and of crisis alike, the Soviet Navy had aircraft patrolling, whose huge fuselages were crammed with electronic detection gear. Out at sea they had electronic intelligence trawlers. Every resource they possessed would be in service now.

'No way,' Clifford said abruptly. 'There's only one circumstance detailed in our orders for taking that risk and it's if the mission gets aborted and this ship has to pull them out. No, Major. Peterson would say the same himself. On no account break silence. He'll guess there's been a word change and go to Red if he doesn't find them at the bay, just take him a while longer. Judging from the maps and photographs it should be a fast hump anyway. Not above three hours, I'd say.'

'Well, you're the boss.'

'Was the boss.' For a moment Clifford let bitterness at his misfortune show. 'You know something, Colonel Peterson could be right to take over. I guess Howard Smith was a bad choice for this operation. There's something too gung-ho about that guy's character and they should have realized it at Fort Bragg. Same time, Peterson himself was always telling us not to think of this in terms of rushing in to win the big medal and yet that's just exactly what he's doing right now.' He looked at his ankle and shook his head morosely. 'Jesus, what luck. I could kill that bastard Russian.'

'We're damn lucky he didn't kill us. In fact we're not precisely doing the safest thing on earth hanging around here for three days.'

Clifford sighed. 'Then we'd better start fishing. Wonder what kind of a medal they give for catching cod.'

The shore of the bay was a desolate crescent, strewn with flattish chunks of ice, the sea a dully rippled reflection of the clouds, until they merged in the distance. Peterson scoured the landscape through his binoculars yet again, noting the dark areas of snow-free ground, the occasional hump of rock, the grey sea itself. The cloud was gaining a discernible outline, instead of being like a cotton-wool shroud. He was carrying his skis and trying to walk on the bare patches to avoid leaving footprints. Nothing else moved.

'They're sure as hell not here,' he muttered to himself, 'and the goddamn fog's lifting.'

He felt extremely isolated. Somewhere on this cold coastline was the raiding force, somewhere at sea was the *Northern Light*. If he couldn't find one, he was going to have to contact the other. Svalbard was no place to be out in alone at the best of times, even though the M-16 was a defence against most animal hazards. He eased the pack off his shoulders, dumped it on the ground and sat down, determined to think this out logically.

If there had been a word change, then surely Weston would have signalled it and he would have been notified, right up to the last moment in the DC-9. But there had been no message. Could the main party have been intercepted? It was a possibility he had to consider, especially in view of the incident with the Russian destroyer. On the other hand, he ought logically to hump the eleven clicks to the other RV on the assumption that the long-range communications had somehow failed. But 11 kilometres could take him three hours. He'd already gone one hour in what looked to be the wrong direction. Suppose they thought a word change *had* been received? How long would they wait before sending men out to search for him? His original orders had specified one hour. As far as they knew he could be lying up there on the drop zone with a busted leg, sounding his emergency bleep and hoping to Christ they'd home in on it. There was always a screw-up when communications failed, like when his own patrol had shot up another from B Company around Hill 147 that night. Always had been screw-ups, always would be. What was worse, they'd be looking in the wrong area and exposing themselves to no purpose.

As if to emphasize his fears, the whine of an aircraft's engines sounded overhead, out of sight above the cloud.

'One thing's for sure,' he told himself. 'You have to get the hell out of this place.' For the first time since Admiral King had approved the plan, he wondered if his parachuting in hadn't been too bright after all. He and the team could lose half a day trying to find each other. He came to a reluctant decision. He was going to have to check if there had been a change.

He unstowed the small PRC-77 radio from his pack, squatted beside it, extended the aerial, tuned to the frequency and listened in the headphones, then pressed the transmission switch for a second. A buzzing click sounded in his ears. The set was live and working. He pondered the briefest possible message, pressed the switch again, and sent it.

'India Whisky Charlie. This is Dixie Six. Interrogative Yellow over.'

There was a long pause. When the answer came he could hear the snap in Clifford's voice.

'Six negative Red Out.' There could be no doubt of the feelings on board the *Northern Light*. The concluding word 'Out' had an angry curtness which made it clear they expected radio silence.

Peterson closed down the set and straightened up. At least Smith would now know he was on the way. He set off, glad to be starting towards the mountains and the mist. In the mountains there would be snowdrifts, overhangs of rock, long shadows cast by the low sun, whereas down here there was no concealment.

Once inland he put on his skis again, though the snow was thin and cloggily granular, like wet sugar. The skis fitted to his boots with a metal clip over the toe and a strap around the heel, known as a Kandahar binding. The boots themselves were 'ski-march' ones, acquired from the British commandos. He had Norwegian *taladd* or cloth anklet coverings over them to stop his sweat freezing in the leather. One thing he had learned years ago about winter warfare was that other nations' equipment could sometimes be better than your own. The only thing he had not been able to acquire was the Norwegians' instinctive mastery of bad conditions, the ability of men whose mothers had first fixed skis to their feet at the age of three or four. None the less he swung determinedly across the sticky snow, pushing on each ski in turn so as to glide a yard or two forward on the other. An observer would have thought it a clumsy imitation of the *langrenn* gait, but with 90 pounds' weight on his back there could be nothing classic about his progress.

Periodically he halted, checking his compass and the map. After nearly an hour the slope leading to the mountain ridge became faintly visible again under the low cloud. A few hundred metres farther on he saw something which made him stop and instantly drop flat, discovering once again that though the binding made it possible to thrust one's skis out to each side, it was still awkward. Sure, he thought, they do it in ski and shoot competitions, but not with all this damn gear on. He lay panting from the exertion, groped for his binoculars and cautiously studied the snowscape. There was movement ahead of him, a group of dark shapes half hidden in the mist.

Several long minutes passed, then suddenly he burst into

suppressed laughter as the shapes became more distinct: these were friends, at least potentially.

Slowly the reindeer came nearer, heading north across his route, presumably looking for snow-free pasture near the shore. They were a small species, Peterson noticed, grey-white in colour with dark antlers and short legs. Round their eyes was a circle of near-black hair, for all the world as if they had king-sized hangovers. Comic, but maybe useful. He knew there were ten thousand in Svalbard, and the Nordenskioldland area which the raiding party had to cross was one of their principal habitats.

He had an untested theory about the reindeer. Their body weight couldn't be so different to that of a man, their thick coat would keep in warmth much as his own arctic clothing did. In fact, as seen on an aerial infrared photograph recording sources of heat, a Svalbard reindeer might be indistinguishable from a man. So, just possibly, a Russian photographic interpreter, analysing the product of the sorties being flown high above the cloud, might have several thousand tiny spots to confuse with those made by the team. Might. The idea had occurred to him too late to have it verified.

As soon as the reindeer were past he struggled to his feet and continued. After a further hour, when despite his fitness his shoulders were starting to ache, the water of the Bellsund came in sight. He halted and took stock. The unending cloud was now a clearly defined ceiling to the landscape, and the fog beneath had cleared, so that although the mountains were hidden, the far shore was visible as a horizontal line.

The routine from here on was standard. He would advance until he was challenged. He carried on another click until, scarcely 50 metres ahead of him, a white-clad figure rose beside a hump of snow, a sub-machine-gun a black outline in his hands, his face shadowed by the white hood of his parka.

'Northern,' said the sentry very quietly.

'Star.'

The sentry beckoned him forward, then saluted when he recognized Peterson.

'Glad to see you, sir.' It was Millar, the British commando, carrying out his final duty as organizer of the beach party. 'Captain Smith's a couple of Ks towards the mountain, sir. It was getting pretty exposed down here, what with the fog clearing and all.' He rubbed his mittened hands together vigorously. 'Cold too, sitting like a bloody Eskimo without an igloo.'

Peterson smiled, recalling Millar's gift for expressing himself.

'You all OK?' he asked.

Millar nodded. 'No problems, sir.'

'Let's hit the trail, then.'

Millar led, following the tracks left by the main party, skiing with a relaxed confidence which reminded Peterson again that the commando mountain leaders were just about the equals of the best Norwegians in snow. Some difficult compromises would have to be made in balancing his own troopers' carefully nurtured pride and aggressiveness against the need to make the most of all the expertise available. He wondered how, precisely, Howard Smith had figured on getting the outfit together.

Peterson's altimeter needle was nudging 200 feet and their track had changed to a herringbone pattern up the slope when they reached the patrol base. Smith had used the tent sheets, buttoned as usual in sets of seven and supported by ski poles, to make low shelters, well spaced apart and surrounded by snow walls. Over them were draped white camouflage nets. They certainly wouldn't defeat aerial photography, but they had been pitched in the misty base of the cloud and close up against the mountain. Short of digging snow holes, Smith had made the most of the position.

'Good to see you,' Peterson said, and crawled in under the flap of one of the shelters. Mydland was squatting and

Burckhardt, the radioman, was crouched melting a cook tin full of snow on a small spirit stove.

'Coffee in ten minutes, sir,' Burckhardt said cheerfully.

'Reckoned we'd better hole up and wait, Colonel,' Smith explained. 'We had Johnson and Trevinski at your drop zone. They only quit when we all heard your transmission.' There was definite condemnation in his nasal New England voice, as if to say 'the whole goddamn world must have heard you'.

'My orders were to RV at Yellow, skipper.' In trying to tone down his anger Peterson slipped into the old Marine slang for a captain. 'Given the circumstances those drop zone coordinates could never be firm. No word change reached Bodo. Who authorized it?'

'We signalled, sir.' Burckhardt looked up from the stove. 'Got what we thought was an answer, but it was badly distorted. I guess it's this weather, sir. Hell of a lot of interference in these latitudes. I've fixed a pretty damn good aerial, though. If it don't work, nothing will.'

'Have you notified this checkpoint yet?'

Smith answered, his expression unrepentant, 'We will do now you're here.'

'Right. After that we should get off our asses. We have to find a way past the mountains and the glacier. The more distance we put between ourselves and the LZ the better. Then we can retrieve those stores and get organized.' Peterson turned to Mydland. 'You have any views on the route we planned, Captain?'

Mydland spread out his map as well as the confined space allowed. Parallel with the west coast there ran a high range of mountains. The closeness of the contour lines emphasized how precipitous they were. On the other side lay a 3-kilometre-wide glacier, which descended to a small inlet off the Van Mijenfjord. Once past this inlet, at its narrowest only 700 metres across, there was an easy way to the Reindalen.

'That inlet may still be frozen over,' he said. 'If it is, then we could save ourselves a lot of time. If it isn't, we shall have to work our way up until we can cross the glacier.'

'I'd go for the mountain route from the start, keep inland and out of sight,' said Smith. 'That was my plan.' He had argued with Mydland about the route. 'OK, it could be an ass-kicker, but it's secure in this weather.' And if you hadn't come busting in, he was thinking, everything would have been fine and we'd have gained several hours.

Peterson studied the map, as though he didn't have both possibilities permanently imprinted on his mind. 'It's a question of speed versus risk. The minimum height we can cross the mountain is a thousand feet.'

'We can reach the inlet in two hours,' Mydland suggested. 'The pass could take a day. And we have to find it in thick mist.'

'We'll follow the easy route and stay as high up the slope as we can.' Peterson made his decision. He believed in listening to his subordinates' ideas, but keeping them well short of TV quiz-programme style. He looked at the radioman. 'You'd better send that signal, Burckhardt. I guess we're all qualified coffee-makers here.'

Burckhardt wormed his way out and Peterson concentrated on the order of march. In terms both of map reading and trail breaking, this first stage would be simple. He wanted to save the Norwegian for later, when the navigation became difficult and the snow deep.

'You be point man on this hump, will you, Smith? We'll saddle up just as soon as we've had the coffee and Burckhardt's through. Jesus, it's nearly eleven already.' Peterson began rummaging in his pack for his rations. The airline meal he had had on the DC-9 seemed part of a forgotten life.

Twenty minutes later the radioman returned, failure evident in his face. 'Can't raise them, sir,' he said dejectedly. 'There's interference like a crowd at a ball game. You want I should go on trying?'

Peterson grunted, sipping the last of his coffee, reflecting on priorities. They had two more reporting points, the final one immediately before the assault. It wasn't likely that Admiral King had any word change to make at this stage. What mattered was to get to the objective.

'You can close the set down, Burckhardt,' he said. 'We'll shoot the breeze with them at the next checkpoint, OK? Right now I want to get this show on the road.'

Buckhardt retreated, thinking that the Colonel didn't change. 'Expedite' had been his favourite word right through the training and it still was. Jesus, why did I volunteer for this piece of action? If the operation remained out of contact, there was only one guy would carry the can, namely Corporal Joseph C. Burckhardt Jr of Merrill, Wisconsin. Hell, he'd read all about high Arctic radio problems. He should have seen this coming.

The young Russian controller, whose arrogance had seemed to increase day by day as spring thawed into summer, was gazing down through the long windows. Below on the tarmac a border police guard of honour was forming into three ranks under the direction of a sergeant major.

'Well,' Folvik declared, full of resentment yet undemonstrative as ever, 'I'm off now. You got rid of me at last.'

'What a pantomime!' The Russian spoke ostensibly to himself, but loud enough for Folvik to hear and catch the contempt in his voice. In Norwegian he said, 'So your Governor lost his nerve.'

'On the contrary, he's being recalled for consultations.'

'And you? Does your King want your advice too?'

Folvik clenched his fingernails into his palms. Even though he had long ago realized that it was deliberate policy for Soviet officials to speak as though Norwegians were answerable to their monarch rather than to the elected government, the ploy still angered him.

'I have been posted to a new job,' he said.

'You are a lucky man.'

Folvik stared at him. How much did he know? Could he

have guessed that intelligence was filtering back to Norway? In truth, Folvik did not know why he had been ordered to leave. If he was being removed for his own safety, why was Annie left behind?

'This is no longer a good place for Norwegians.'

'Nor for Russians!'

'What do you mean?' The young controller was instantly alert.

'Nothing.' Folvik tried to shrug off the indiscretion, thinking nervously that he would never make a real spy. Since Sven's death the tension had been near boiling point. Last night Annie had received instructions about recruiting the football team to trigger a riot if need arose. He had searched out Lars-Erik in the Store Messe and approached him – successfully. Something big was happening all right. To avoid the Russian's eyes he turned to the long windows.

Beyond the parked DC-9, aboard which he would so soon leave for ever, stretched the black swathe of the runway and beyond that were scattered patches of snow on the slope of the Plata Berget, rearing up into the cloud. A stranger would hardly guess the mountain existed. He tried to imagine the radar he had never seen, rotating slowly up there in the deep snow, a lonely symbol of Russian conquest, but found it hard to visualize, even though a radar was the one thing he and the other controllers had most wanted for Longyearbyen in happier times. He still felt the Plata Berget was his mountain.

His attention passed to the interior of the control tower, the long array of radios under the windows, the gleaming chrome of the Negretti and Zambra barometer, the clock now indicating 0838 GMT. He noted their details lovingly, these instruments of his private world, and his eye fell on the cabinet at the back of the room where the kettle and jars of coffee still stood. He had quite forgotten his Nescafé. Why the hell should this arrogant young communist have it, with his boasting about the Komsomols and local Party leadership, and his acid remarks? Folvik walked across and picked up the jar, thrust it in his coat pocket and strode out. The action relieved his feelings a little.

The junior staff were escorted aboard the DC-9 first. Walking out, Folvik saw the sturdy figure of Makarov, evidently keeping an eye both on the guard of honour and on a crowd of local people gathered behind the barriers, waving Norwegian flags. The high wire fence ran from the tower to the administration building, a distance of perhaps 10 metres, and the people were jam-packed behind it. Folvik recognized Annie from her long hair streaming out beneath a jaunty fur hat. She saw him, shouted and twirled her flag high. He raised his hand in salute, assailed by guilt at leaving the burden of responsibility to her alone. She blew him a kiss and called out 'Safe flight.' He was glad when one of the two blue-uniformed Norwegian police on duty signed to him to hurry up. It was a relief that the police were remaining.

From a window seat, he peered out through the circle of Perspex, trying to catch a last glimpse of Annie. She was a wonderful woman, an inspiration. The guard of honour crashed to attention as the Sysselmann came out. Makarov was at his side, marching stiff as a puppet in his greatcoat. Folvik could hear the crowd defiantly singing the Norwegian national anthem. Up above he saw the pale face of that odious young controller pressed to the glass.

The guard presented arms and the Sysselmann quickly inspected them, his expression as wooden as Makarov's. There was no sign of General Stolypin. That must be deliberate, calculated to demonstrate that of the two theoretically equal governors, the Soviet one was the more equal. Folvik had learned in the past few months what enormous importance the Russians attached to rank and protocol.

The Sysselmann was welcomed aboard the plane by the captain, hydraulics whined and the rear door clunked shut. As he fastened his seat belt in response to the usual announcements, Folvik felt the hard bulge of the coffee jar in his pocket. He pulled it out. His initials painted on the lid stared up at him. FF.

Frederik Folvik, the son for whom his mother had such dreams, who saw his name in lights, he thought miserably. Oh hell, what a failure, but at least I tried.

As the DC-9 rumbled down the runway, gathering speed, and rose punctually into the cloud at 1000 GMT, eleven o'clock local time, the confusion of memories and hopes in his mind became almost unbearable, bringing the sting of tears into his eyes, until one idea dominated his brain with mournful clarity. He was not cut out to be a hero, and some power greater than himself, whether divine or human, had appreciated the fact.

'So Prebensen is safely out of the way? Excellent. Sit down, Comrade Colonel.'

General Stolypin's slit of a mouth puckered into a momentary smile and he removed his steel-rimmed glasses, as if in a gesture of frankness. He knew perfectly well that the DC-9 had left on time but wanted to convey a touch of beneficence to Makarov by implying that the departure was an achievement, which it was, though not the Colonel's. However, a pat on the back was in order, given the nature of the next task. The time had come for Stolypin to distance himself from the trouble that now seemed imminent.

Obediently Makarov sat down facing the wide black desk which had so recently been Prebensen's. Through the window he could see that the weather was deteriorating again. Fog was rapidly obscuring the fjord.

'If the Norwegians think recalling Prebensen will add to the diplomatic pressure on us, they are totally wrong. On the contrary it will enable me to exercise a far closer control of the local population.'

This was no surprise. Makarov had a fair idea of what Stolypin's instructions from the Politburo must be, namely to lose no chance of turning the screw tighter on the Norwegians. But he noted warily that Stolypin said 'me', not 'us'.

'Stricter control is necessary because the American action we were warned about may have begun.' Stolypin picked up a teleprinter message lying on his desk and replaced the glasses on the bridge of his nose with a frown of irritation. 'Our electronic intelligence picked up two radio transmissions in the Bellsund area early this morning. Both were very short. Unfortunately a fix was only obtained on the second. It showed a position close to the coast. At 77 degrees 46 minutes north, 12 degrees 29 minutes east.'

Makarov's glance strayed automatically to the large map on the wall.

'There are no NATO warships in that area, yet the transmissions were on a short-range VHF frequency.' Stolypin held out the piece of paper. 'The conclusions are obvious. I have ordered immediate infra-red photographic coverage.'

Makarov stood up, took the message and scanned it quickly. The intelligence analysts at Murmansk inferred from the use of the words 'yellow' and 'red' that a landing was either imminent or had already taken place, since the United States forces normally used colours to denote beach landing zones. He moved across to the map. Ever since his training as a young officer, he found that scanning heights and river lines and valleys stimulated his tactical thinking. If the Americans were mad enough to try some kind of raid, then the Bellsund would be a good place to come ashore unnoticed, though a helicopter could place them much closer to either of the radar sites. Happily, he could rely on the listening devices he had installed, after the capture of those two spies, to pick up an enemy helicopter.

Of course, he suddenly thought – the spies! They could have been reconnoitring routes, pre-positioning supplies even. He checked the map again. The trappers' hut was 25 kilometres from the ice-free part of the Bellsund. He must set up an ambush there immediately. Damn this fog.

'What is puzzling you, Comrade Colonel?' Stolypin was quick to sense a mood.

'A minor consideration only, Comrade General.' He wasn't going to reveal this hunch. If he was wrong, he could be ridiculed. If he was right, Stolypin would steal the credit. 'I presume you will want my defensive plans for the radars activated?'

'At the Isfjord radio site, yes.'

'Not up here?' Makarov was astonished. The only operational military target so far established in the whole of Svalbard was the radar on the Plata Berget.

'No. Not up here. In my view the mountain constitutes its own defence. More important, the early-warning radar will be at the Isfjord radio. It is the obvious target.' Stolypin's tone became drier and more condescending. 'There is something you can now be told. The ship carrying the main radar masts sailed from Murmansk last night. For political reasons, a merchant vessel is being used. The masts are almost completely assembled and so have to be deck cargo. We must assume they have been identified by American satellite photography and that an attempt might be made to prevent their erection.' He paused. 'In fact we already knew that military action was under way.'

Makarov turned back to the map. The mountain range near the coast ran much closer to the shore at the northern end where the Isfjord radio station stood isolated on a flat promontory. Any raiding party approaching overland would be entering an ever narrowing funnel between the mountains and the sea.

'They would be rats in a trap,' he said. 'We could cut off their retreat to the south. It's a perfect killing ground.'

'We will want to take them alive, Comrade Colonel. Alive.' In his imagination Stolypin visualized a group of captured Americans being paraded before the press and TV cameras in Moscow. What propaganda!

'With respect, Comrade General – ' Makarov saw the delight of anticipation in Stolypin's expression but could not believe it justified, ' – would even this American leader take such a risk?'

'A previous president risked and found humiliation in Iran.' Four months ago Stolypin would have been prepared to philosophize at length with Makarov. Now, having cast him as the potential scapegoat, he was inclined to be brief, but he enjoyed analysing the failures of the West. 'The only recent president who understood matters in a world context was Nixon, and he allowed himself to be destroyed. Most Americans see individual problems as resolvable by an appropriate individual response, especially when they are politicians seeking re-election. The democratic system itself prevents their planning in the great depth of time which we enjoy. So they might not relate our establishment of radar in Svalbard to strategic developments on the Chinese border, let alone to the encouragement of revolutionary movements in Latin America. They might think simply of NATO being de-stabilized and of a direct threat to themselves, with a limited action seeming the correct answer to our limited advance.' Stolypin stood up behind his desk, signalling the end of the interview. 'In any case, Comrade Colonel, we know that they have been planning such action. Your orders are to prepare the defence of the Isfjord radar.' He smiled sardonically. 'A job fit for a hero, eh? Perhaps you will win another medal.'

Makarov clicked to attention and departed for his own office, where he telephoned for a Spetsnaz patrol to stand by at Barentsburg. He had lunch brought to his desk, consumed it hastily while he drafted orders, and then allowed himself the luxury of writing a letter to his wife. Only to her could he unburden his feelings.

'My darling Svetlana,' he wrote. 'There is little I can put on paper. This morning the Norwegian Governor left. You will have read about that in *Pravda*, I am sure. Personally I respected him. He was an honest man attempting the impossible.'

He paused, absent-mindedly chewing the end of the pen, sifting phrases in his head. The probability of this letter being censored was high. 'Today I was allocated new tasks.

"Fit for a hero," the General told me in that well-known manner of his.'

Again Makarov hesitated. He wanted to convey his fears, so that if he did end up disgraced she would know it was not his fault, but he had to do it obliquely, so she would read between the lines. 'In the critical situation now developing here, who can tell whether I shall survive?' He grunted. Yes, a censor would assume he meant survive in battle. She would know it meant survival in his career.

'It is impossible for security reasons to explain further. However, I have always been lucky. I was lucky at the Academy, lucky at the bridge in Angola. You have always said, "You were born lucky, Viki," and you are usually right. Look how lucky I was to marry you! So probably there is no cause to worry. If the present situation becomes easier I will immediately ask permission for you to come here. It's not so bad in the summer.

'I send my deepest love to you and to Alexei. When does he take his examinations?

'Your loving husband.'

He signed it neatly, as befitted a former engineer, then went out to the helicopter waiting to take him to Barentsburg. Surveying the weather again he reckoned it should be possible to fly along the fjord, skimming the surface. But reconnaissance inland would be out of the question. The brief flight confirmed this and on his arrival he sent the patrol to be dropped on the shore of the Van Mijenfjord. The men could ski to the hut.

'Please sit down, Sandra. You don't mind the Christian name, I trust. We have met before after all, quite old friends, eh?'

Phillpotts's beaky face creased into an avuncular welcome as he gestured her to an armchair.

'Not the most elegant room, I'm afraid.' His bony hands deprecated the small office allocated for his visit and he meticulously eased the trousers of his blue suit as he sat down himself. 'Trouble with this place, they only recognize military rank. Walk in without any and no one knows which way to look. Thank God they've at least provided some coffee.' While he kept up the casual patter, his eyes were busy comparing impressions with those he had recorded a year ago and re-read this morning.

Tarted herself up a bit, that'll be Julia's influence, he thought, or else the boyfriend's, and surely she used not to be a blonde? Still the suburban tennis party type, though, faintly gawky, self-conscious, jolly hockey sticks. So what the hell does this Kraut see in her? As he handed her a cup of coffee he tried to examine the ring and saw the glint of diamonds but could not properly distinguish the individual stones. Damn it, I'm getting old. Long-sighted as an owl. But not too old to know when a girl has what it takes and, by God, this one hasn't. So why the diamonds? Why marriage?

'Milk and sugar?' He offered the jug and bowl.

'No sugar, thank you.' She half smiled, uneasy at being scrutinized, then forced herself not to worry. He's absurd with his old-fashioned language and his cavalry moustache and his regimental tie. Uncle Philly. Everyone in the office laughs about his annual visits. She bet he never even dreamed that Julia was called 'the bicycle' behind her back. They were all agreed that Uncle Philly was the original innocent. Probably didn't even know how to spell the word 'sex'. Her affair with Erich had made her feel very worldly. She sipped the coffee and decided attack was the best form of defence. After all, he was just a re-employed civil servant, a 'retread'.

'So you're back again.' She tried to make it sound challenging.

'Yes,' he murmured, divining the trend of her thoughts. 'Have to come occasionally. Too many people here are more concerned with their own status than with the enemy.'

'Well, we aren't at war, are we?'

'You think not? I'd say we're pretty close. And can't there be an enemy within?' He let the phrase hang in the air a moment. 'So, I hear you're engaged. Doesn't do to congratulate the girl, eh? But I'm sure you're thrilled. Pretty ring, too, by the look of it. Who's the lucky fellow?'

'Erich's a writer.' She instinctively bolstered the image, avoiding the word 'reporter'. 'He represents a magazine in Munich.'

'Ah, yes. *Aktuelle Nachrichten*, isn't it?'

'How did you know?' Her temporary self-confidence was fading. She hadn't liked that reference to an enemy within. She had told Erich nothing he couldn't have discovered by himself, nothing specific at all. Even last night, when he had been more casual than usual and only asked about Svalbard in passing, all she had revealed was that the crunch would come this weekend. She didn't know any details; she had simply heard General Anderson saying so.

'How did I know?' Phillpotts inquired benignly, not missing the confusion in her voice. 'Oh, I think Julia mentioned it when we were talking earlier. She's something of a gossip, I fear.'

'But she doesn't know Erich!' What was Julia doing meddling, damn her? Sandra's cheeks reddened with anger.

'I gathered you all met at the flying club once.'

That was true. Julia had been out at Grimbergen one day, showing off her latest acquisition, an Italian Air Force major who she claimed was an aerobatic virtuoso. Boudoir aerobatics, more likely.

'Is Erich a good pilot?'

'Oh yes! Marvellous. We went to Deauville two weekends ago. That's where we spotted a ring just like this.' She glanced down at her finger lovingly, though there was also a shadow of disquiet in her expression. 'It is lovely, isn't it?'

'Very nice indeed.' Phillpotts put on his spectacles and leaned forward. 'Quite delightful. Do you hope to go on working after you're married?'

'Not if he gets the job he's hoping for in Berlin. His family are there, at least his mother is.'

'And his father? We would like to know his father's address. If you were to carry on here, that is. Purely a formality, of course.' Phillpotts coughed gently as though embarrassed at his own probing. 'You know the system.'

'I don't really know about his father.' Sandra felt sudden panic. What did she know about Erich, save the purely physical? She knew he had a brown mole on his shoulder, she knew the hard bulge of his arm muscles, the way he breathed when he was asleep. Of his past she knew practically nothing. He never spoke of it.

'Perhaps you'd ask him some time.' Phillpotts's tone became a trifle less avuncular. Everything he had learned in this interview so far tied in disquietingly with the Munich police report. Erich Braun was effectively a man without a past. Most telling of all, his tax history began much too recently, last year in fact. A twenty-seven-year-old West German who was fit enough to fly an aircraft would certainly not have escaped the draft. During national service he would have acquired a tax number. He must have spun an amazingly good story to some tax office clerk to get around that most characteristic of all German requirements, for everything to be in order. Well, *Alles in Ordnung* was not a term the Munich authorities were likely to apply to this particular gentleman's record any longer, thanks to these inquiries. Not that any of it proved he was an East German agent. But the magazine was communist-funded and the presumption was inescapable.

'Does he cover NATO for *Aktuelle Nachrichten*, by the way?'

'Yes.'

'Doesn't get much out of the Press Office, I imagine.'

'I really don't know.' She was firmly on her guard now.

'And he asks you occasionally? For help, I mean.'

'It would be odd if he didn't, wouldn't it?'

'None the less, you don't tell him anything, I hope.'

'He usually seems to know things before I do.' That was true. Erich was uncannily perceptive. He had appreciated the importance of Svalbard before she had even heard of the place.

'Such as?'

'Well . . .' She was beginning to flounder. 'Well, he knew all about Polar Express.'

'So you do discuss things?'

'Not seriously. I mean not things that matter.' She was being drawn into admissions she didn't want to make.

Phillpotts regarded her through his spectacles with the concerned gaze of a family doctor about to reveal an unpleasant diagnosis.

'Sandra. Please think carefully before you answer this question. Has Erich ever asked you about NATO military planning?'

She sat silent. Her greatest fear was that they might question Erich. If she let him in for any kind of trouble, he would be furious. He might even leave her. She had convinced herself that her life depended on being faithful to his needs. But Phillpotts would never understand that. The moment had come when she had to dissemble.

'No,' she said, her voice not completely steady. 'He did ask once, but I didn't tell him. Most of it's Greek to me anyway.'

'Asked you about what?' He had recognized her directness as that of a direct lie. She could never be even an adequate secretary to General Anderson if she didn't understand military jargon.

She faltered again, awkward under his scrutiny, and he saw her fear. There was acute danger in putting words in her mouth. If he made the wrong suggestion she would accept it and exploit it. He took a risk, based on the *Aktuelle Nachrichten*'s coverage.

'He's been writing about the Svalbard crisis recently, hasn't he?'

She nodded, not trusting herself to speak.

'For the last time, have you told Erich anything about our Svalbard planning?'

'No. How could I? It doesn't go through us. I told you Erich knows far more about Svalbard than I do.' She let her outburst develop its own momentum. 'Why do you go on asking? Can't you see, all I want to do is get married and leave this place!' She began to cry. She had told the essential truth at last and it hurt much more than the truth he wanted.

'Here, my dear.' He pulled a carefully folded white handkerchief from his breast pocket and handed it to her. 'I didn't wish to upset you.' He waited while she wiped her eyes, then rose courteously, the old-fashioned gentleman once more. 'I'm sorry I had to ask so many questions. You'll be here tomorrow, won't you? Good. Just in case there's anything else.'

He left the possibility with her and ushered her out unhurriedly. He wanted her to sleep on the idea, preferably to sleep alone. She was lying all the way. That was obvious. The problem was, he knew even less about NATO's Svalbard planning than she did.

Forty minutes later Phillpotts was concluding a taut conversation with Brigadier Curtis, the Chief of Staff, who had been extremely reluctant to believe that a nice girl like Sandra could conceivably be mixed up with a potential East German spy. At heart Curtis was precisely the kindly, somewhat old-fashioned military gentleman that Phillpotts merely pretended to be. In a less critical situation Phillpotts would have derived some amusement from the role reversal, but this was not a moment for laughter.

'She's lying in her teeth, Brigadier,' he concluded aggressively. 'And I'm damn certain she's told her boyfriend something connected with Svalbard. What can it be?'

He could see fears of breaching security written all over Brigadier Curtis's face.

'For heaven's sake, man. If you don't trust me, whom are you going to trust?'

'There are some top-secret Anglo-American plans, of

which General Anderson and I have few exact details. No need for us to know.'

'Is she privy to them?'

Curtis struggled with his memory, then his face went white under the freckles. He was remembering last week, when the General had called her in and she had waited patiently by the door as they finished their conversation. Anderson's words could be quite meaningful if coupled with the requests he had made from NATO countries to counter possible Soviet retaliation.

'She could know that limited military action was imminent,' he said heavily. 'I imagine the Soviets know about the possible naval blockade. Most of the NATO measures have been intended for their consumption.' Briefly he explained what Sandra had overheard.

'And the military action?'

'Is clandestine and has already begun.' Curtis almost added that he had considered it extremely risky from the start.

'Then we can't wait until tomorrow.' Phillpotts was briskly matter of fact, as though everything was now clear. 'I'll have to break her today. This afternoon. Give her a few hours to ponder this morning's session.'

'Will she talk?'

'In a very high pitch. When she learns that her boyfriend has also been sleeping with Julia.'

'You're not serious. The little bitch!' Curtis shook his head. 'God, these women.'

'A loyal bitch at least.' There was one overriding moral principle in Phillpotts's world: loyalty excused most things. 'It was Julia who tipped us off that Erich might be a bad 'un. You may yet have a lot to thank her for. The truth will be painful, but it should persuade Sandra to confess. She's jealous enough of Julia as it is.'

'That's a pretty mean approach, isn't it?' Curtis was genuinely shocked.

'Depends on your priorities, Brigadier. Are one traitorous girl's illusions worth the life of a single soldier?'

'No.' Curtis swallowed uneasily. The implications of all this were sinking in rapidly and they were frightening. 'No, definitely not. Christ. The General will go berserk. Why in hell does it have to be a Brit who betrays us?'

Phillpotts looked at him coolly. 'Because the majority of your secretaries are British. If NATO paid American rates it could as easily have been an American.' He smiled for the first time. 'Sex has become remarkably internationalized of late. As for your General, he should know better than to gossip in front of the staff.'

Mydland surveyed the inlet with dismay, noting streaks where the covering of snow had melted and the ice was either gone or perilously thin. The Van Mijenfjord usually stayed frozen well into June, due to its mouth being largely blocked off from warm sea currents by a long island. He had hoped the inlet would still be solid, even though it was close to the sea.

'Can we chance it?' Peterson asked, leaning on his ski poles. The going so far had been easy, a skate, and his instinct told him to hurry on, select what stores they needed from the cache, and get up into the safety of the mountains. 'I want to expedite this move. We'll lose one hell of a lot of time if we can't ski across.'

The inlet, seeming so innocent on the map, lay at the end of a glacial valley, its shores a rock-strewn moraine. The kilometre of moraine on each side would be difficult going, but quick compared with crossing the glacier.

'Too dangerous, I think.' Mydland shook his head. 'My advice is to climb the mountain alongside the glacier and cross high up.'

The whole group looked inland. The end of the glacier was a good 5 kilometres off, just visible beneath the cloud. Even at this distance they could see that it was a jagged wall of tumbling ice blocks. The mountains enclosing it rose like citadels into the cloud.

'What a bastard!' Smith commented. 'What a goddamn bastard.'

'Looks like you're going to get the mountain hike you wanted earlier,' Peterson said grimly. Because there were so many glaciers in Svalbard he'd been boning up on the subject. This was what most of Europe and North America had been like in the Ice Age. The glacier facing them was lying in the mountain hollow like a deep river of snow, compacted by pressure into striated ice, moving 1½ to 2 feet every twenty-four hours, fissuring into treacherous crevasses, scouring the surface ground to bring down the debris which formed the moraine. It was damn near impossible to plan a route across Svalbard without hitting several glaciers. 'Well,' he said confidently, 'let's just beat the hell out of this problem, shall we? Guess we ought to rope up for the climb.' He turned to Millar. 'You mind breaking the trail, Sergeant?'

'Any time, sir.' The wiry commando's cheerfulness never deserted him.

They set off, edging their way up and along the precipitous slope with painful slowness on their snowshoes and carrying their skis. But at least the mountainside was better than the moraine. At 240 feet by Peterson's altimeter they reached the cloudbase, clambering on up into a misty white wilderness, barely able to see 50 metres ahead. Mydland reckoned they could cross the glacier at around 500 feet, but the 6-kilometre distance took three hours. They had just reached it, finding to their relief only a tiny crevasse between the glacier and the mountainside, when a distant noise made Peterson whistle the signal to halt.

Somewhere out of sight below a helicopter was flying. The sound was muffled by the snow. Peterson stood stock still, listening. Damn it, he could hear two. For sure. 'What d'you make of it?' he asked Mydland quietly.

'Perhaps because of this weather they cannot go direct to Svea and are flying along the fjord.' The Norwegian hesitated. 'If I may suggest, Colonel, it might be prudent for me to go ahead and reconnoitre when we approach the trappers' hut. The border police did occupy it for a time.'

'Your people told me that was three months ago.' Surely to God they hadn't been holding back intelligence!

'True. I only suggest being prudent.'

They crossed the glacier successfully, skiing in single file. It was noticeably colder in the cloud and the contours of the snow had no definition in the diffused light. It was hard to tell if they were going upwards or downwards, until a gradual increase in momentum warned of a descending slope. But they trekked steadily on, each man lost in his own particular thoughts, each developing his own individual aches. The soles of Peterson's feet began to burn inside his boots and he regretted their newness: he had intended to break them in during his time at Bodo. Well, he told himself, they're going to be old friends in three days' time.

During a halt Trevinski tackled him on the question of using a pulk; two of these lightweight sleds were with the stores in the cache. The Corporal's black moustache was beaded with ice crystals. He raised his dark goggles to look at Peterson.

'You know, sir, this goddamn Dragon may not be the weight of the TOW but it's no chicken either. A pulk would make a lot of sense.'

'You see yourself hauling a pulk up these mountains, Trevinski? You must be kidding.'

'We can always bury the thing again if the going gets tough, sir.'

'Well, I'll consider it.' He still didn't like the idea. A pulk needed two men to pull it. But he had no illusions about the effect carrying the missile had on a man's performance. He was humping one of the rounds for the Dragon himself now. As far as he was concerned, rank made no difference to the distribution of loads.

Shortly after five Mydland considered they were within a kilometre of the hut. They halted and the men wrapped themselves in their tent sheets and prepared to eat dry rations while the Norwegian probed forward. He left most of his equipment with them, slinging a sniper rifle over his back and setting off in a long and cautious traverse down the valley. They had been keeping in the cloud all the way, but no higher than necessary, and as he descended the visibility improved rapidly. The mist began breaking up into patchy areas of haze. He became able to see 50 yards, then 80, then 150, then much more. The log hut came into view, a dark brown rectangle on the whiteness, its roof thick with snow. He stopped automatically and dropped flat to survey the valley below.

The rock by which they had cached the stores was not so easy to identify from above. He was fairly sure that he had recognized it when to his horror the door of the hut opened and a white-camouflaged Russian soldier appeared, a sub-machine-gun slung over his shoulder. Mydland pressed himself down, digging his chin into the snow, praying the man would not look upwards or that if he did he would not see the less white shadow in the mist.

The Russian deposited a pair of skis on the snow outside the hut and bent down, quickly fitting them to his boots. He glanced round, then set off in the opposite direction. Mydland eased out his binoculars and followed the soldier until he stopped at what appeared to be a distant mound of snow. Two tiny figures rose from the mound as he reached it. Mydland swore silently, stowed away the binoculars, eased off his skis and began to crawl back up into the concealment of the cloud, his thoughts churning. God, but the Russians could be persistent! They must have been sending patrols to this area ever since Folvik first reported it. If Peterson wanted the stores he was going to have to fight for them. If they were still there.

The aerial reconnaissance report, predigested by the intelligence analysts at Murmansk, reached Makarov in the early evening of Wednesday. Meanwhile the Spetsnaz patrol had found nothing at the trappers' hut, but were in position for an ambush. He was now ensconced in the operations room at the Barentsburg helicopter base, away from the mining town on the small promontory of Heerodden.

The base was far better equipped than anything yet available at Longyearbyen. Enlarged without Norwegian permission in 1978, it had hard standing for twenty helicopters, a short cement runway and a range of buildings, including hangars and a drab brown two-storey administration block, surmounted by a small control tower. From the circle of windows, on a good day, he would be able to see the radar on the Plata Berget through binoculars, while the Isfjord radio station was only 15 kilometres in the other direction. But the real asset was the operations room below, from which surreptitious contacts had long been organized between the original five 'civilian' helicopters and the Soviet Northern Fleet. The Heerodden base was designed to support just such an operation as Makarov had to mount; it had the communications facilities; above all, its security was uncontaminated by Norwegian civilians. It could have been the key to the takeover; as it was it would serve the defence.

Makarov had waited impatiently for the report, knowing that within hours the weather would destroy the chance of immediate action. By the time the long-range aircraft had finished patrolling to and fro, far out of sight above the cloud, and the film had been developed and interpreted, the fog had rolled in over the whole west coast. The Isfjord radio station was reporting 50 metres visibility at sea level as Makarov began marking the information on his maps. The information was startling.

'Then the General was correct,' exclaimed the Spetsnaz captain with him. 'There is a group ashore. And a support ship.'

'It seems so.' Makarov finished the neat red signs denoting an enemy vessel 14 kilometres off the coast and a force of platoon size ashore, moving north. 'Estimated strength twenty men,' he read out aloud. 'No vehicles.'

'In other words, a commando-type raid, Comrade Colonel.'

'That's what the experts think.' Makarov skimmed through the rest of the signal. 'There are also 107 unidentifiable heat sources away from known settlements of which investigation is recommended, a group of men with vehicles near Sveagruva – those must be the border police – and our own radar on the Plata Berget. I had better have a word with the General.'

The quickest communication with Longyearbyen was via the Isfjord radio. Makarov went up to the control tower and returned a few minutes later.

'The Navy is intercepting the ship,' he said. 'A boarding party will very quickly establish what it's been up to. We are to prepare an ambush for the suspected enemy patrol, sending a force immediately by boat to cut off its retreat. No helicopter is to fly within 10 kilometres for fear of alerting them. High altitude photographic reconnaissance will be carried out as soon as the weather clears to identify the other 107.'

'Those can't all be men.'

'Polar bears, probably,' said Makarov. 'Now, let's get down to detailed planning.' He would have to replace the troops at the hut with police, damn it. A single company of Spetsnaz was not enough for all the tasks he had.

Sandra narrowly avoided collisions twice driving home. She couldn't concentrate, her hands shook on the wheel, she felt sick and apprehensive. After the second near miss a traffic policeman waved her down and was cautioning her when he noticed how ill she looked and asked if she wanted to be taken to hospital. She insisted that she could get back all right, and drove on slowly. She was as much in a state of shock as if she had been involved in an accident. The afternoon interview with Phillpotts had hacked at the roots of her being.

When she reached the apartment she took three codeine tablets, forced herself to drink hot milk and tried to calm down. Phillpotts had attacked her relationship with Erich, accusing her straight out of passing him information.

'Even if I did unintentionally mention something,' she had admitted eventually, 'he won't misuse it. I can trust him.'

'You think so?' Phillpotts had said harshly. 'Then ask him where he spent the night of the DPC meeting.'

'I don't understand.'

'While you were dining with Brigadier Curtis, he was climbing into bed with your colleague Julia. That, Sandra, is how far you can trust him.'

She had stared at Phillpotts, too horrified to speak, then burst into tears and sworn that it couldn't be true. When her outburst was over, he had asked if she would think again about Erich, and she had passionately refused to do any such thing. Before he let her go, and he had done so with patent reluctance, Phillpotts had handed her a visiting card, with a Brussels number neatly added. 'For your own good,' he had said. 'For your own future, challenge him. If you need help, phone me. At any time. The middle of the night, if need be.' Then he had looked her straight in the eyes and said quietly, 'I'm afraid your fiancé may not be everything he seems. It is very important indeed that you remember anything you might have inadvertently told him. I promise that no one will blame you.'

She sat half crouched in an armchair, Phillpotts's phrases tearing at her. How could Erich have betrayed her? She fondled the ring. It simply was not believable. Then she remembered the morning after the DPC meeting. Julia had been particularly brazen and she had caught a glance from her full of sly curiosity. At the time she had wondered about that glance. Now the implications made her too unnerved to drink the hot milk, even though she knew it would be soothing, and she poured herself a large whisky instead. Erich would be here at eight. What was she going to do?

8

Eight o'clock came and Sandra was still terrified. She tried to pull herself together and begin preparing the meal. When the bell gave its familiar three rings to indicate Erich was on the way, she was so tense that she cut herself with the vegetable knife.

'My God, what a day!' He came in, threw his coat over a chair and kissed her seemingly as an afterthought. Then he noticed her agitation. 'Are you all right, my darling?' he asked.

She stood there, the apron tied round her waist, her eyes welling with tears.

'What's the matter?'

She shook her head mutely and moved away when he tried to take her in his arms.

'Why do you not tell me?' Irritation mingled with concern in his expression. He was tired and he had worries of his own. A message from Diederichs in Munich had warned that the police had been asking questions, a lot of questions. The last thing he wanted now was a scene. He wanted to be fed and to sleep.

She caught the note of intolerance in his voice and reacted, screwing herself up to speak. 'Are you faithful to me, Erich?' She stood there, beseeching him to say yes, to say he loved her, to give her reassurance.

'Faithful?' he echoed. It flashed through his mind that she knew about Julia. 'Of course I am. We are engaged.'

It wasn't enough and he realized the inadequacy a fraction too late.

'What were you doing on the night of the DPC meeting?' She was becoming uncharacteristically shrill. 'Erich. I must know. Are you playing about with other women?'

'Darling, it's you I'm in love with. You must believe me.' He tried to take her hand, swearing to avenge himself on whatever bastard had told her.

'Don't touch me!' The falsity of his words, his expression, everything confirmed it, and for the first time in her life she became hysterical. 'Go away! Leave me alone! I never want to see you again. Ever, ever, ever!' She began shouting, suddenly pounding her fists on his head and chest. 'I hate you!'

'Shut up!' He slammed his right hand against her face, spinning her across the floor until her legs caught a chair and she fell on the little tiered drinks trolley, which collapsed and splintered beneath her weight. Glasses shattered, the bottle of whisky fell and broke open, spirit gurgling on to the carpet. He looked down at her angrily. There was only one way to treat a hysterical woman in his view. He was in a mood to beat someone up anyway.

She lay there in the cracked debris of the trolley, disbelief shading into fury. Never in her whole sheltered life had a man hit her. 'So the security people were right,' she spat at him. 'You did go to bed with her. Oh God, I have been a fool.' She struggled to get up.

'The security. . . ?' He didn't attempt to help her. All today's problems clarified into the realization that he had been shopped. 'What have you been telling the security?' It occurred to him that he ought to kill her. But that could wreck his chances of escape.

'Nothing.' She was on her feet again. 'Nothing except that I loved you.' She leaned down and picked up the whisky bottle from the floor, initially out of disgust at the stain it was making. When it was actually in her hand and she saw its jagged neck, her bitterness suddenly found an outlet. She swung round and lunged at him, shouting, 'How dare you hit me?'

He reacted fast, but not fast enough. The broken bottle sliced the palm of his outstretched right hand. He swore, not yet feeling much pain, cast around for a weapon, and tried to pick up a chair. The agony of gripping it was terrible. She advanced on him, holding the bloodstained bottle like a weapon.

'Get out,' she said. 'I'll give you one last chance. Get out and leave me.'

He gave her a contemptuous smile and left.

When he had gone, she stood in the middle of her room, totally shell-shocked for several minutes. Then she went into the bedroom and found Phillpotts's visiting card. She sat on the bed, gazing at the telephone, trying to decide. So the old man had been right. His words came back to her. 'I'm afraid your fiancé may not be everything he seems.' She had assumed Phillpotts referred to Erich being unfaithful. Could he have meant more, that she had been doubly betrayed? Was Erich a spy, not a journalist? But he *was* a journalist. She had seen his office and his articles in the magazine. Yet the word 'security' had hit him like a blow. 'What have you been telling the security?' And when she had told him to go, he had left with no word of regret, no pleading for a second chance.

She fetched another bottle of whisky from the store cupboard and drank some neat. Several times she almost lifted the telephone and each time a fear that she had still not thought things through prevented her calling Phillpotts. Eventually she admitted to herself what had happened. Erich had never loved her. Her only ever fiancé was gone, lost irretrievably, and she had been used, horribly used.

Slowly, as she accepted this idea, she progressed to considering the future. Whatever Phillpotts promised, there was seldom immunity for betraying secrets to an enemy. And that was what she must have done. For love, but how would that help? She would lose her job, her few friends, this apartment. She would be tried and jailed, with terrible publicity, and her parents would want to die of shame.

She never did call Phillpotts. At three in the morning she took the rest of the packet of codeine, ran the bath, got in and summoning a last, half-numbed effort of will, cut her left wrist with a razor blade and watched the blood run like a red fern into the water.

Weston shook himself out of a half doze, shivered and wiped an arc of the wheelhouse window clear with his palm. Through the smeared moisture he could see that snow was falling outside, settling on the *Northern Light*'s rail and on the heap of nets lying on the foredeck, but melting on the deck itself. He checked the time, angry with himself at having fallen asleep. It was 5.22 and a new day, Thursday. The trawler was hove to, its engine closed down, the radar off. Clifford and Paulsen, the Norwegian mate, were asleep in the cabin, stretched out on the bunks. They planned to cast the nets after breakfast.

Periodically during the night whoever was on watch had turned on the radar for long enough to establish the *Northern Light*'s position. She was drifting very slowly north with the current. This was a normal fisherman's routine and when not emitting signals herself the ship should attract little attention. Weston poured himself a mug of coffee from the thermos flask, then switched on the echo sounder. A zigzag line began to be traced across its paper. They were in 30 fathoms of water. He looked at the map. The depth tallied with where Paulsen wanted to fish. Later they would use the Atlas Fischfinder to locate the shoals. He turned on the radar. The beam of light began to rotate, making the screen glow where it revealed the outline of coast. But in the opposite direction, out to sea, the blobs showed the radar echo of ships. Two ships. He studied them. Each time the yellow beam went round, re-illuminating the blob, they were a fraction closer. Timing their progress against the concentric rings denoting distance on the screen he began calculating. Whatever the ships were, they were proceeding with caution at barely 8 knots. By the time he had watched for a few minutes he knew both were converging on the *Northern Light*.

Though he had no means of identifying the ships, Weston knew he must presume they were hostile. He compared the map with the radar screen, scratching himself idly through the thick wool of his sweater. He felt badly in need of a bath and wished, not for the first time in his career, that the military did not always take action at dawn. When he had come to a decision, he went back to the cabin and roused the two others.

'Bad news, I'm afraid,' he told Clifford. 'I think the Russkies have rumbled us.'

'Can we run for it?'

'At 11 knots? We're boxed in. The only place to run to would be the Isfjord.'

'Could we scuttle her and go ashore?'

'What do you mean, "scuttle"?' asked the mate, in his accented English. 'Do you mean sink the *Northern Light*? If that is so, I cannot agree.'

'I wasn't asking you,' Clifford said roughly.

The Norwegian stiffened with anger. His private orders from the *Northern Light*'s owner were to save her from unnecessary damage, whatever might be said about compensation.

'Don't worry,' Weston intervened soothingly. 'We'd be worse off ashore than we would on board. All we can do is signal Bodo that we're being intercepted, then start fishing and try to brazen it out.' He chuckled. 'Paulsen, you can tell the Soviets they're not allowed to sink an innocent fishing boat. We've got about an hour and a half to hide the incriminating evidence.'

They heard the low whine of the destroyer's ventilation machinery sounding eerily through the snow-muffled gloom before they saw the ship itself. The radar had already told them one was closing in while the other had cut off any retreat southwards. Weston glanced quickly around the *Northern Light* for one final time. The trawl ropes stretched taut over the side, among them a single unobtrusive cable from which hung a waterproof bag containing the military radios. He had put a short seaman's jacket on over his white sweater. The mate had no need of disguise. He wore a typical Norwegian fur-lined hat, the ear flaps up over the top, and his stained clothing emphasized his profession. Clifford lay inside the cabin, his ankle conspicuously bandaged.

The only aspect of the ship Weston could not fully hide was the hold. In preparation for the trip he had fitted up black tarpaulin shielding for the bunks, which when fixed in position provided a hold within the hold. Full of fish, it would almost certainly pass inspection; empty the space was obviously much smaller than it ought to be. Someone looking down from above might conceivably not notice. But if the door from the engine room was opened, the game would be up.

'We have to keep them away from that bulkhead door,' Weston told the Norwegian.

'I will try.' Paulsen felt their chances were small, especially in view of their treatment at the last encounter with the Soviet Navy.

As he spoke, a dark grey hull took shape in the falling snow, the rearing line of a destroyer's bow. There was a churning of water as she went astern to stop and then she lay there, 30 yards away, dwarfing the *Northern Light*. A figure in a greatcoat and high peaked hat came out from the bridge and raised a loudhailer.

'Stop your engines!' The order was in Norwegian, each word articulated, yet sounding more distant than the ship.

'Why?' Paulsen cupped his hands round his mouth to shout back.

'We are coming aboard. Stop your engines.'

There was no alternative. Inside the wheelhouse Weston

switched off the two diesels, whose steady throbbing died with a splutter. Then he and Paulsen stood and watched as a small inflatable craft was lowered and six men, armed with sub-machine-guns, swarmed down a rope ladder into it. One seaman pulled on a cord to start the outboard, they heard him rev it up and the inflatable came spurting across the narrow stretch of water.

The Russians were taking no chances. Two sailors stayed in the boat, covering Weston and Paulsen, while a third threw a grappling iron up to the *Northern Light*'s rail. Ropes were secured and the men climbed swiftly aboard, an officer leading.

'Who is the captain?' demanded the officer, who spoke passable Norwegian. He stood on the deck, a revolver in his right hand, young and truculent. Without being ordered the sailors took up positions at both ends of the *Northern Light*.

'He is,' Weston said, indicating Paulsen and speculating on how many officers of the Soviet Northern Fleet were trained as linguists. This one's command of the language was quite as good as his own.

'What are you doing in these waters?' the lieutenant demanded.

'Fishing,' said Paulsen shortly, pointing to the ropes and beginning to feel more angry than frightened. 'What do you think?'

'You are in the territorial waters of Svalbard.'

'Yes. Norwegian territorial waters.' The mate barred the entrance to the wheelhouse.

'All shipping has been warned to avoid this area. We shall search your ship.' The lieutenant raised his revolver. 'Stand aside.'

Weston heard him questioning Clifford, who was unable to reply adequately, and from then on the search party opened every locker and examined every map. Inevitably they discovered the false interior of the hold and the dismantled Zodiac in its bag.

The lieutenant prodded Paulsen down into the engine room at gunpoint.

'What is this for? What is concealed here?'

'Nothing is concealed.'

But the attempt to brazen it out was foredoomed. The hold was clearly furnished for sleeping.

The lieutenant spoke briefly on a walkie talkie to the destroyer, then faced the two men. 'You and your ship are under arrest. You will haul in the nets and proceed to Longyearbyen.'

As they heaved on the ropes, watched closely by the Russians, Weston decided he must try to cut the cord holding the radio with his clasp knife, while pretending to free a fouled line. He succeeded in tangling part of the nets, swore loudly in his best Norwegian, then knelt by the rail, leaning over the side, tugging at the rope. Keeping his back to the Russian nearest him, and his gaze on the nets, he reached for his knife and unhurriedly hacked at the cord.

'Stop!' A bullet whined wide of Weston's head and splashed in the water.

As the lieutenant fired, Weston felt the last strand of the cord part. He pulled himself backward and upright, the clasp knife still open.

'What's the matter?' he demanded.

The Russian came forward, leaned down and picked up the cord, examining the severed end and realizing that it was not necessarily a part of the trawl.

'Get inside,' he said, furious at being outwitted. 'Touch anything else and I will shoot to kill.'

Weston shrugged his shoulders, not trusting himself to reply fluently, and went into the cabin. At least the most incriminating piece of evidence was out of the way. Through the windows he saw that the nets were now being dragged aboard by the Russians. The few fish already caught were left squirming and wriggling in the mesh, their bodies flashing like dull silver until they were buried in the mound of netting.

Two hours later Paulsen was edging the *Northern Light* past the Isfjord radio station, invisible in the snowstorm, bumping chunks of ice as they sailed towards captivity.

Glaring orange neon illuminated the platforms and tracks, loudspeakers played echoingly soulful music. Outside it was already morning, but no daylight penetrated the station buried beneath the terminal building. It was as though the management of Brussels Central wanted their passengers to feel even more subdued than the early hour would already make them.

There was a whirring sound as the figures on a train indicator flicked over. Erich Braun spun round as if threatened, as if some unscheduled departure might be announced. His bandaged hand made him feel conspicuous. He had spent the night destroying documents and notebooks until his landlady had bawled up the stairs at him to stop the noise and for God's sake let her sleep. Then, with his few important possessions in a suitcase, he had driven to the office, cleared his desk and left a note on the typewriter. There at least the night watchman was accustomed to journalists' irregular hours. Even so Erich was frightened in case he had aroused suspicion and he watched the indicator's black and white numbers apprehensively until they stopped.

'Koln 0640.'

It was a *Rapide*. He took a deep breath of relief. In an hour and a half he would be across the frontier, shortly after nine in Cologne. By then his car would have collected its first parking ticket, but it could remain like that all day on a meter without attracting serious attention. By the time the police found it, even assuming they started to search, he would have left Cologne/Bonn airport on the next stage of his journey.

A draught of air and a rumbling announced the arrival of the express. The carriages made a groaning noise as they rounded the curve, the wheels hissing on the rails, the brakes squealing. He picked up his suitcase and glanced cautiously both ways along the platform. The guard was already out and walking along towards him, an absurdly young man with long hair tumbling out from beneath his blue cap. Decadent bastards, he thought as he opened the nearest door and jumped in.

Four minutes later the red second hand of the platform clock jerked past the vertical, the young guard's whistle shrilled and the train began to move. Braun leaned back in his seat, holding up a newspaper to shield his face. He hoped the coffee trolley would come round soon.

For a few moments Peterson thought he was entombed in an avalanche, trapped in a pocket of air, only alive by a miracle. Keep calm, he told himself, keep still, don't panic. In the same instant he woke fully and the dim interior of the snow hole took shape around him. He was on a narrow shelf of hard-packed snow, separated from his sleeping bag by only a thin mat. He rubbed his eyes and wriggled his aching shoulders. The reindeer skins Norwegians used to have for insulation were better, but far too bulky.

'Jesus!' he exclaimed to himself as he remembered yesterday and how they had been forced to abandon the cache of stores. A lot of decisions had followed from that, mostly unpleasant, starting with a dogged eight-hour hump up into the safety of the mountains, which had left the whole team exhausted. And when it was over and they had dug these holes out of a sloping drift, they had still been unable to raise Bodo on the radio. What a mother of a day.

He looked across at Burckhardt, still lying asleep on a similar shelf less than two feet away. Between them was a roughly dug trench to catch the cold air, while the light of a single candle revealed the roof of the snow hole arching irregularly above them. The candle not only gave warmth. As long as it stayed lit they had enough air coming in through the vent hole which he had poked through to the outside with his ski pole.

He checked the time. Eight minutes after eight. Zero eight zero eight on Thursday morning, he repeated to himself, a curious coincidence of figures. For a few more moments he savoured the luxury of lying down, staying warm, not having to move. Like Sunday mornings. Memories of Nancy overwhelmed him, her dark hair against the pillow, the musky scent when he kissed her neck and her slow, sleepy response to his own sudden passion. 'Tom, darling,' she would always protest, 'why do you wake up so *early*?' He said a silent prayer for her and for his own return, then forced his mind on to immediate problems. The loss of the cached stores was not a catastrophe: they had been intended as a backstop of provisions in case the team needed to stay holed up in the mountains longer than planned. The pulks, the extra food, the ammunition buried down there were all bennies, a bonus over and above what the team had brought in with it. OK, so Trevinski wouldn't get his pulk. The main point was that they had to complete this operation within the original time scale or else run short of supplies. That meant the weather would be even more crucial. He eased himself out of the sleeping bag, extracted his boots, kept there in a polythene bag to stop the leather freezing, pulled them on and crawled to the entrance.

The way in and out of the snowhole was by a short tunnel with a vertical exit into the drift. This was camouflaged by a foldable cover that Peterson had devised himself. When he closed the hole he had laden this cover with snow and lowered it into place after him, supporting it with his ski sticks. Now he reversed the process, pushing it upwards like a trapdoor and emerging cautiously. You could survive bitter cold and be well concealed in a snow hole, but it made him feel like a hunted animal, sniffing the air for danger at the entrance to its lair.

Snowflakes drifted against his face, soft and large, melting swiftly on his skin. None the less he could see some hundred or two yards down the slope of the Reindalen valley. He wondered what the visibility was like down at sea level. Bad or good, the eight-hour hump had been doubly worthwhile. They had got far enough up the valley to remain in the protection of the cloud and now these heavy flakes were covering their tracks. More than that, though he had laid out the position here himself, he could hardly tell the location of the other snow holes. In fact he'd taken the risk of relying on concealment and not ordering any sentry roster. He called for Smith softly.

'Yes, sir.' The muffled reply was followed by a patch of snow collapsing. Smith's face appeared in the opening, framed by his white-hooded parka.

Both men clambered out into the open. Smith bent over, fixed on his snowshoes and padded across with ungainly steps.

'Your turn to navigate,' Peterson said. He wanted to give Smith responsibility, which wasn't easy at this stage. 'We'll check the route with the Norwegians. You eaten yet?'

'Sure thing.'

Peterson nodded, noticing that Smith had also managed to shave. For himself he preferred to keep what little water Burckhardt would have time to boil up for the essential hot drinks. In any case, a growing beard gave a little protection to one's face. But Smith was definitely one of the 'they died with their boots clean' kind.

'Good,' he commented, genuinely appreciative of the Captain's readiness. 'I'll just snatch some chow myself and give orders in half an hour.'

By the time he had crawled back into the snow hole, Burckhardt had lit the tiny gas burner and begun melting snow. Preparing food was a slow business, even the freeze-dried kind which, if necessary, you could eat unreconstituted. As Peterson waited, hunched up in the confined space, with the plastic-coated map on his knees, he made his decisions for the day, scribbling notes on a small pad. One thing more than any other worried him.

'Burckhardt,' he said at length, 'you reckon this radio interference will get any less as we go higher?'

'Shouldn't make too much difference, sir. Not on HF.' Burckhardt didn't look round. He was concentrating on making the most of the small quantity of boiling water in the cook tin. 'I guess things'll change with the weather. You want I should try again?'

'I'd like that.'

Laying out the aerial on the snow would require hardly five minutes, even allowing for taking compass bearings, and the coded Morse message was already stored in the set. All Burckhardt had to do was press the transmitter button and wait for a reply. He could do that while listening to the orders, because the set would also store incoming 'bursts'.

When they had eaten and were ready to move, the group stood round, balancing on their ski poles, Mydland intent not to misunderstand anything the Colonel said.

Peterson looked them over, glad yet again that he was leading the raid himself instead of merely being the intermediary. He had come to know and value these men, and all the history he could remember told him the same thing about military operations: when it came to the crunch, the soldiers on the ground carried the can for the politicians, from the president down. He knew he couldn't lead from the front for ever, but unexpectedly doing so now made him feel great and he slipped easily into his old, confident way of delivering the routine order sequence in a personal style.

'No need underlining what our situation is and all that,' he began. 'There's enemy patrolling and we'll be using this bad weather and the terrain to keep clear of them. No need either to think it's more than routine.' He paused before the next item.

And if it is more, Millar thought, watching Peterson carefully, then we're up the creek without a paddle and no mistake.

'Friendly forces,' Peterson continued. 'None, except at sea, and, as you all know, we have a slight communications problem.' He caught the flicker of apprehension on Burckhardt's face as he looked up from the radio and instantly countered it. 'Purely temporary and it won't affect coordination of our withdrawal because we'll be on the VHF set then. Correct, Burckhardt?'

'All the way, sir.'

'I guess there's something too goddamn wrong to put right if you don't all know the mission. I want to be at our final laying-up point tomorrow evening, so we can get good and ready to take out that radar early Saturday. Before dawn would be if there was a dawn. That gives us around thirty-six hours to cover maybe thirty-five clicks on the map. Case that sounds like a skate, just remember those thirty-five will turn into fifty or more working up valleys and over passes.' He looked at Mydland. 'I want to expedite this move and I'm going to take the most direct route possible. I'll show you on the map in a moment.'

He turned back to the others, noticing how their faces were already becoming weathered by exertion and exposure. In another two days they'd feel they had lived in these clothes for ever, their bodies would smell rank and they'd hate the sweat drying in their crotches and armpits. They'd have to be careful of that too. As they climbed higher and the temperature fell, any sweat would freeze as soon as they stopped to rest.

'Now our progress could be pretty slow at times where the going's tough and I'll be changing the point man every couple of hours. I'd like you, Sergeant – ' he nodded at Millar ' – to trailbreak first with Captain Mydland behind you. Captain Smith will navigate. I don't expect to see any enemy, but if this weather clears and we do, well, I guess you all know the SOPs. And in this bad visibility we'll keep close together, no more than two or three yards apart, right?' He continued quickly through the few administrative and signals matters, emphasizing that the loss of the cache was no kind of disaster, then asked for questions.

Millar reminded them to change ski wax for the fresh

snow, suggesting yellow klister, and Peterson then gathered them round closer to explain his preferred route.

' 'Stead of following along east up the Reindalen, we're going to strike north and find the pass to the top of that Colesdalen valley.' He let them find it on their own maps, fumbling the glossy folded sheets with gloved hands. 'Then around behind that real high bastard, the Nordenskioldfjellet, which dominates our target.'

'You do know, sir,' Mydland said cautiously, 'that there is a Russian settlement on the Colesdalen?'

'A disused mine, right? Well, Captain, we'll pass nine kilometres inland of that.'

'Sure seems the fastest route, Colonel,' Smith cut in. 'We lost some time; we should catch it up whatever way we can so there's plenty in hand before the assault.'

'Correct,' Peterson said. Maybe he and Howard Smith did think along the same lines after all. There were occasions to expend a team's energy, others to conserve it. Mydland said nothing more, so Peterson gave the order to move in ten minutes and joined Burckhardt, who had remained with his radio.

'Do you have any contact yet?'

'Negative, sir. No reply.'

'Better saddle up then.'

When Burckhardt had gathered up the aerial and the whole team had waxed their skis and adjusted their loads, Howard Smith took over giving directions.

'Remember, keep right in the one track. We don't want to give anything away we don't have to.'

He told Millar to head off and they swung away easily, traversing across the slope to the valley's centre before beginning the long climb up to the first pass, a straggled line of ghostly figures, soon swallowed up by the cloud and falling snow.

The Belgian gendarme in his dark blue uniform stood aside

after he had rung the bell of Sandra's apartment three times. 'Open the door,' he instructed the elderly concierge.

'She's never in late mornings, except at weekends,' the woman protested. 'Is she in trouble or something?'

'The sooner you open up, the sooner we'll know,' the gendarme remarked tartly. He didn't want any nonsense, not with these two senior British officers standing waiting and the headquarters backing their request, though personally he thought the fuss was absurd: the girl was only late for work, after all.

The woman fiddled with her keys and eventually swung the door back. The gendarme entered first, as befitted the representative of authority, and called out loudly, 'Anyone here?'

'Looks as though there's been a struggle,' Phillpotts commented, noticing the broken drinks trolley and moving towards it.

'Touch nothing, please!' The gendarme noted the disorder before going through to the bedroom. A moment later he returned, his face paler. 'She's in the bath,' he said in a subdued voice. Although he had seen many corpses, death still distressed him.

Phillpotts thrust his way through. Sandra lay on her back in water so crimson-stained that it largely hid her nakedness. Her head lolled against the side of the bath, her hair was dank, her eyes closed. On the side of the bath was the razor blade, neatly put there before she died.

'I never thought she was the kind,' Phillpotts murmured to himself. He dipped a finger in the water. It was cold. He touched her shoulder. Her body was stiff, the flesh almost as white as ivory with a bluish tinge. She must have been dead a good few hours, how many he was not expert enough to judge. He returned to the living room, wondering if her address book was in a drawer or her handbag, then realized he didn't need it. He rustled through the phone directory and rang the number listed for *Aktuelle Nachrichten*.

'Can I speak to Erich Braun, please?'

'I regret you have missed him.' The man answering sounded Belgian. 'He is not coming back today.'

'It's urgent.'

'He has gone to Munich. He left a note.'

'By car?' Phillpotts was already calculating distances to the frontiers with Germany, France and Holland.

'That I don't know. Would you like the magazine's Munich number?'

Phillpotts took it, rang off and called the gendarme. 'Can you get me through to your headquarters, please?'

'You think Braun did it?' Brigadier Curtis asked incredulously.

Phillpotts shook his head. 'No. But he was probably here. Two whisky glasses on the table, blood on that broken bottle. If we don't catch him at the airport or the frontier, we never will. One thing is sure. He won't be on his way to Munich now.'

The gendarme interrupted saying his inspector was on the line and while Phillpotts explained his reasons for wanting to arrest Braun, Curtis wandered round the little apartment, though he avoided the bathroom. There was something peculiarly sad about the abandoned possessions of the dead; make-up that would never be used again, an almost empty scent bottle, a stained fragment of tissue. 'Poor kid,' he thought, half expecting to find a suicide note. What could that bastard Braun have said to her? Not to mention Phillpotts. Sandra had been a good and likable secretary, who had responded eagerly to his wife's friendship. He found it hard to believe she could be a deliberate traitor, even in an organization as riddled with intrigue as NATO. True, there had been others. But Ursula had been a straight spy, while Valerie, who had secretly joined the Greenpeace movement, and then defected to East Germany, was widely held to have been psychologically maladjusted. They were before his time anyway, in the 1970s, part of history. Could Sandra really be the third secretary on the Military Committee's staff to have gone bad on them in a decade? Surely that numbly serene body in the bath could only be the outcome of extreme unhappiness, of rejection by her lover?

'Must we assume she betrayed us?' he asked, when the telephone conversation was over.

'Unfortunately, yes.' Phillpotts never quite lost his old-fashioned air, even at his most brusque. 'I told you that before, Brigadier. I misjudged her reaction, though. Thought if I pushed, she'd fall our way.'

'It must have been hard for her. She was very much in love.'

'Besotted, you mean.' Phillpotts did not disguise his contempt. 'Our American friends have a phrase for this situation. They call it "the honey trap". Usually it's men who are seduced, that's the only difference.'

As Curtis was driven back to Evère in his staff car he began summing up the implications of Sandra's still unproven guilt. They ran to a considerable number, from the immediate necessity of withdrawing the Svalbard raiding party to an undermining of Anglo-American relations at the very moment when the Alliance most needed to display solidarity. General Anderson was not going to like any of it, especially since his own loose talk was involved.

In the operations room at Admiral King's headquarters the whole eastern Atlantic and Arctic situation was displayed. The combined results of every form of intelligence gathering, from satellites high above the earth and sensors on the seabed to individual sightings by trawlers and reports from straightforward, old fashioned spies, all helped to pinpoint the vessels of the Soviet Northern Fleet. They and the ships of the hastily assembled NATO Task Force were manoeuvring in a vast natural arena of water, enclosed to the north by the Arctic pack ice, to the west by Greenland and Iceland and to the east by the Russian islands of Novaja Zemlja. Around this expanse of sea squatted the air bases: the Soviet concentration of 350 combat planes controlled from Archangel and the Kola Peninsula; the lone Norwegian airfields of the far north, strung scantily around the fjord-indented coastline; the American base on Iceland; the RAF stations in Scotland.

Admiral King stood, bareheaded and thoughtful, evaluating the strategic problems. To aid his concentration he had lit a cigar, something he rarely did in the mornings, let alone this early. He was silently reminded of the time by the range of clocks showing the different local time zones. Here in Norfolk, Virginia, it was 08.58, in Svalbard 14.58. Although it was Thursday morning here, by the time he had obtained political agreement to any major move, it would be nearing midnight where Peterson was and little more than twenty-four hours before the team's attack on the radar was due.

Distance too might be crucial. All those map locations showed one thing conclusively. To avoid what those goddamn diplomats called 'provoking' the Soviets, few US Navy ships were in the Barents Sea east of Svalbard and Bear Island. The bulk were 'exercising' to the west and – if hostilities resulted from what he planned – that left one big question mark hanging over the sea area just east of Greenland known as the Lena Trough. The lines of seabed sensors stretching out from the coast of Norway did not go that far because it was too deep. The Trough was one route for submarines to approach the North Pole under the ice. He was unable to tell what Soviet subs might be lurking there now, effectively in the rear of his main force.

He drew on his cigar, noting where Norway's five frigates and three corvettes were stationed. Only four of the five were operational but, happily for his purpose, he had one with the NATO Standing Force at his immediate disposal. The Norwegians figured prominently in the scheme he was devising. It was all a question of time and distance. At least, thank God, time and distance were finite and calculable parameters, not like human reactions. Despite King's self-controlled, if set, expression, he was almost in a state of shock. He could scarcely believe that some two-bit limey secretary had betrayed Peterson's mission for the sake of being laid. It was grotesque, insane, completely mind-blowing. Yet he had to act on that assumption, and it hurt him worse than if his wife had been unfaithful or his son jailed for a drug offence. The British were always producing traitors, damn it. Philby, Blake, Blunt, all those others. Despite his habitual calm, Admiral King was unable to keep the bitterness at bay when talking to his British NATO deputy.

'That bitch has put us on the spot, Frank,' he said angrily. 'The *Northern Light*'s been intercepted. It must be Peterson's turn next. We have to pull the team out, and fast.'

'Is that the chief priority?' Rear-Admiral Frank Williams had watched the Svalbard involvement grow since that long-ago day when he had first met Peterson in the office upstairs. Recent events had heavily underlined the delicacy of timing required for a clandestine raiding party to do its job and be extracted. The theory of following destruction of the radar with a message on the presidential hot-line telex to Moscow saying that this was a non-public warning of the US government's unwillingness to see Svalbard militarized was feasible. Just. But it relied on the Delta Force team achieving total surprise.

'What happens if we lose the team?' he insisted.

'A lot of us are going to be in retirement. There's no way we can hush up the loss of those men, even if the Russians don't put them on show in Red Square.' That was the guts of the free world's problem, King knew. The Politburo was not answerable to their citizens. The President of the United States was to his. 'It'll be worse than the Teheran rescue.' He turned to Williams, the stress he was under showing in every line of his face. 'All right, Frank, I know the political priority is to stop the establishment of that radar, but a part of it is extricating Peterson.' He paused. 'In your estimation, would the Norwegians take the lead in enforcing a blockade of Svalbard?'

'They damn well should. It's the line they've advocated all along.'

'Then I'm going to ask them to challenge and turn back that Russian merchantman.' King called to one of the staff officers. 'Commander, when's that ship with the radar masts estimated at the Isfjord?'

'Early hours of Saturday, local time, sir. She's making slow progress, probably due to weather.'

'Then we'll shoot for Friday midnight, GMT.' King had made up his mind and forgotten his rancour in the same instant. 'Frank, that gives you thirty-six hours maximum to key up the NATO logistics. Make it so the Norwegians intercept in Svalbard territorial waters and we have a lot of muscle available to back them up. I'm going to Washington, just as soon as I've spoken to the President's office.' He called for the aide again. 'Commander, I want a destroyer detailed immediately to make contact with the raiding team and lift them out. Rules of engagement with the enemy will follow shortly.'

'Yes, sir,' snapped the commander. Like the rest of the staff at SACLANT, he had watched the two fleets' dispositions, weaving about like fencers before a match, and was glad decisions were being made at last.

'And for the record,' King added, 'I am personally authorizing the extraction of the Delta Force team. I sent them in, I'll damn well get them out.'

Peterson might be leading his men into an ambush at any moment.

'Hey, I figure we've hit the pass.' Smith was exultant, though he kept his voice low. Visually the landscape was greyed out, with the powder snow merging confusingly into the fog, but he could feel his skis moving more freely and the altimeter read 740 feet, which was roughly correct. They must have reached the cleft in the mountains which led to the Colesdalen.

'Let's have a halt, then.' Peterson poled himself forward

to join Smith. 'Guess you're right,' he agreed. 'In fact we could be pretty near out of this cloud too.'

It had stopped snowing a couple of hours ago and there was a hazy brightness overhead, though no sign of the sun.

'I'm going farther up with Captain Mydland,' Peterson decided. 'We'll follow our own tracks back. If anything goes wrong, blow the whistle.'

The gradient steepened and the cloud thinned as they climbed. Peterson was reminded of being in an airliner circling Los Angeles once, when smog and cloud blanketed the city and the other planes in the holding pattern were riding through the top of it, their fins as prominent as sharks in a grey sea. A further few minutes spent labouring upwards brought them out into breathtakingly clear air.

Below, as far as they could see, cloud filled the valleys like cotton wool while above, maybe only 500 feet higher, more cloud made a grey blanket stretching to the horizon.

'Hell,' Peterson exclaimed, 'we're the salami in the sandwich, and look at that bastard.'

Straight ahead, across the hidden Colesdalen, there reared up an almost sheer mountain face, a snow-covered wall ribbed with occasional strata of exposed rock, which vanished up into the cloud again.

'The Plata Berget is beyond,' Mydland said. It seemed unnecessary to comment further. They both knew from the map that this wall of rock was part of a formidable, club-shaped bastion of peaks rising to 3500 feet. The haft of the club was a long and equally high ridge on the far side. This ridge terminated at and towered over the Plata Berget. Alongside it was a U-shaped glacier, one arm of which ran down the Longyearbyen valley, stopping a few kilometres short of the settlement.

'Once we are on the glacier,' Mydland remarked calmly, 'we should have no serious problems.' If the map was accurate they could skirt the moraines.

He stopped speaking abruptly, cut short by a familiar noise: the distant clacking of a helicopter's rotor.

'Let's get the hell back down.' Peterson turned his skis as he spoke and immediately started to descend into the protection of the cloud. He snowploughed awkwardly, trying to follow the indentations of their upward path, his heavy kit threatening his balance all the way. If Nancy could see me now, he thought, she'd kill herself laughing. Nancy was a pretty neat skier. He remembered her in a shocking pink outfit, wedeling down a sunlit slope at Aspen with easy grace. Aspen was another world, for sure.

'Let's be clear,' he announced, when they had rejoined the others, 'we have to stay right in this damn cloud just so long as we can. You heard the bird up there. Those guys wouldn't have to be looking to see us. We'll have some chow here and aim to lay up the other side of the Colesdalen, then move again in the early hours.' The sooner they were up alongside that high ridge and able to get well concealed the better. He didn't want to find himself unexpectedly pinned down on a bare mountainside in bright sunshine.

'Should be a nice easy run down.' Smith spoke as if they were at Aspen.

'If it is, then it's the only one we're likely to have.'

'Can only make the most of it, Colonel.' Smith was in determinedly high spirits and his mood proved justified. After they had eaten a basic ration dry, washed down with coffee from the flasks several carried strapped with white masking tape to their web gear, they flexed their arm and leg muscles, relieved themselves in a hole dug with an entrenching tool and set off again. Whereas the 10 kilometres up the pass had taken them six hours, the next 8 downhill were accomplished comfortably in a quarter of that time. The fresh snow sang under their skis and beyond halting when a group of reindeer loomed ahead in the mist, they encountered no obstruction. They stopped finally when Smith realized that the visibility was improving and they must be coming out of the bottom of the cloud. They stood, leaning happily on their ski poles, exhilarated and refreshed.

'Gee,' Trevinski said, echoing what they all felt. 'That was a real skate. I guess it has to be good some of the time.'

'No pain.' Millar echoed the sentiment.

Peterson listened, thinking how even picked men were at heart like any other grunts. As memories of the world faded and last week's steaks and ice cream became as unattainable as riches, they swiftly regained the stoically humorous acceptance of hardship which had been drilled into them since their first days at boot camp. He was relieved that the Briton and the Norwegian seemed cast in the same mould. He was watching them closely, determining exactly how to fit them into the assault they would make in the early hours of Saturday, when the defenders of the radar, if there were any, would be fighting off sleep as well as cold. Around 3 a.m. the body's temperature was at its lowest and so was a man's will to resist. To profit from that, his own team had to be not only confident, which they were, but also rested. He wanted to be over the glacier and into the final laying-up position by 1800 hours tomorrow, giving a good seven hours for sleep and last-minute preparation. It meant driving them hard now. He turned to Mydland. 'You think this weather's staying with us, Captain?'

'There is more wind.' Mydland considered this vital point. 'It's unpredictable here.' He strongly doubted the wisdom of coming into the Colesdalen even though it was a short cut. 'We should keep as far up in the mountains as possible.'

'We'll hump another seven clicks,' Peterson ordered, 'then lay up and catch a few ZZs. Tomorrow could be a busy day.'

'What he means,' Trevinski confided in a cheerful undertone to Millar, 'is that it's going to be an ass-kicker, but colonels aren't supposed to use those words.'

'We closed the trap, Comrade Colonel. We found a herd of these.' The Spetsnaz captain, his rifle slung across his back, pointed inside the helicopter to the carcass of a reindeer. He was young and ambitious and he resented being made a fool of. 'Perhaps we should send a steak to the photoanalysts of the Air Defence Command.' Angry as he was, he did not dare suggest making a similar present to the General, though he knew that the concept of delaying to take the raiding party alive had been Stolypin's.

'You searched the whole area?'

'No sign of movement.'

'None the less, I want a twenty-strong fighting patrol on standby until further notice. Furthermore –' Makarov kept his expression stern ' – don't you know that the reindeer is a protected animal?'

'No, Comrade Colonel, no. I didn't realize that.' The captain's self-assurance was punctured.

'Well, it is, so don't go wasting meat on the Air Defence Command and keep a few steaks for me.' He smiled, hardly able to restrain himself from laughter. 'On this one occasion, this one only, I shall not inform the General. In detail, that is.'

He dismissed the relieved young officer, wishing there was more to laugh about in this damnable situation. Somewhere, concealed in those formidable, cloud-hung mountains, there was an American or British commando unit. There must be. The trawler's equipment admitted of no other explanation. He telexed Archangel reporting the patrol's findings in noncommittal military jargon and requesting further infrared photographic coverage. Then he ordered two of the Mi-6 helicopters to carry out a complete reconnaissance of the coast of Nordenskioldland, going as far inland as weather permitted. Finally, he telephoned General Stolypin and received, as he had anticipated, curt instructions to come to Longyearbyen. 'The bastard can't pin this failure on me,' he told himself, 'but he's clearly going to try.'

War fever and war fear had overwhelmed Britain since the State Department leaked the news that a Soviet merchant ship had slipped out of Murmansk with identifiable early-warning radar masts on her decks. Previously the anti-NATO demonstrations and all the artificial graffiti of slogan-led agitation had been completely unrepresentative of ordinary Britons. True, the recall of reservists had greatly increased the level of concern, but as NATO and Warsaw Pact forces lined the frontiers between East and West, a curious sense of stability had developed, a feeling that this was all for show and no one would actually fight over Svalbard.

Now the Western world was hit by the realization, abrupt as a thunderstorm, that the Soviet Union was deliberately testing its resolve even further and that the East–West confrontation had a new and menacingly short-term focal point. In Britain this realization expressed itself in huge headlines on Thursday morning and acrimonious Parliamentary debate in the afternoon.

Nancy had spent all day in the apartment, praying for a phone call or a message. Nothing came and in the evening she had watched the 5.45 TV news programme with the compulsive attention of an addict, listening with apprehension as the announcer detailed the latest moves in 'the Svalbard Crisis' in his infuriatingly restrained British accent.

'After a day of intense diplomatic activity the Norwegian prime minister declared that Norway would take positive steps to assert her sovereignty in the Svalbard archipelago. At the NATO headquarters in Brussels' – a photograph of the flagpoles outside the long building flashed on to the screen – 'the Defence Planning Commitee, meeting in emergency session, again condemned "the Soviet Union's illegal occupation of Svalbard" and reiterated its full support for Norwegian action, both under Article Four of the North Atlantic Treaty and in accordance with Article Fifty-one of the United Nations Charter.'

Then, dramatically, the programme was interrupted. The

United States Secretary of State was at this moment speaking in Washington. His lean, almost haggard, expression told its own story. Facing an array of microphones he announced, sombrely, that after consultation with its allies, the United States was prepared to back the Norwegian government in every way, including the use of military force if necessary. 'We sincerely hope it won't come to this,' he said, 'but the militarization of this territory cannot be acceptable to the United States nor to any other signatory of the Svalbard Treaty.

'However – ' the Secretary of State had paused, eyeing the unseen cameramen warily ' – the United States recognizes the paramount need to keep all of the Arctic free from Great Power conflict. Accordingly, the President will very shortly be asking the Congress to approve the negotiation, through the United Nations, of a new international treaty to safeguard that territory. Our aim will be, in conjunction we hope with the Soviet Union, to prevent the whole area from military use of any kind and to make it subject to international inspection.'

The moment the hastily prepared map of the Arctic and the North Pole came up on the screen, and the strategic comment began, Nancy realized what this meant. The Soviets were being offered a face-saving formula for backing out of Svalbard at the minimum cost to existing United States facilities. Would they feel inclined to accept it? She prayed they would and feared instinctively that they would not.

Then, surprising herself, she felt a surge of pride that at least her country was standing up to aggression against a smaller ally. So what are the British doing? she wondered. The question was answered a moment later when the now familiar features of the Prime Minister came on the screen, asserting support of Norway in remarkably similar terms to those the Secretary of State had employed. Well, she thought, strangely relieved, at least they've all got their act together, whatever it is. Better still, it sounded like something much bigger than the sort of operation Tom would be on.

General Stolypin read the signal from Moscow yet again. When stripped of rhetoric, it admitted that a blockade was imminent and that the United States Arctic proposals had attracted instant international support. It was now vital to have proof of American duplicity by capturing the men the Norwegian trawler had carried. It should not be necessary to emphasize, the message continued, how shattering a propaganda blow this would be to the imperialist powers. But that blow had to be delivered before the vessel carrying the radar masts reached Svalbard. 'It is your privilege, Comrade General, to serve the cause of the motherland at this crucial hour.'

'In other words,' Stolypin remarked to himself, 'I am on the hook.' The sooner he passed the privilege on, the better. His own greatest ability had always been for survival. He was drafting a reply informing Moscow that Colonel Makarov had been entrusted with the mission of capturing the saboteurs, when a knock on the door announced the Colonel's arrival.

'You've been quick, Viktor Mikhailovitch.' Although he felt no warmth, Stolypin could simulate it if occasion demanded. 'You had a helicopter waiting, eh?'

'A Spetsnaz patrol is at thirty minutes' notice to move, Comrade General. The problem continues to be adverse weather.' Why was Stolypin suddenly sympathetic? He must be worried.

'Good.' Stolypin toyed with a paperknife, twirling it round and round.

Makarov had learned that it was a reflex action of the General's to play with some ornament when he was about to force an issue. Accordingly he watched the thin face with its domed and balding forehead and waited tensely.

'This fiasco with the reindeer has cost us a lot of time.'

'The Spetsnaz commander cannot be blamed.' Makarov was keeping himself well in check. 'I have informed the Air Defence Command of the incorrect infrared interpretation.'

Stolypin grunted imperceptibly. That was quick thinking.

'Good,' he repeated, spinning the tiny glinting knife on the black desk top. 'Meanwhile, I have fresh orders from Moscow. It is Thursday evening. These commandos have already enjoyed a day and a half of liberty since the radio transmission was picked up. The Chairman of the Party himself wants them captured, alive if possible, before Saturday. You have very little time.'

'And if we have not found them by then, Comrade General? Nordenskioldland is a large area.'

'Whether they are alive may then be less important.' Stolypin explained the Soviet Union's dilemma. 'None the less, it is our privilege to defeat this imperialist adventure.' The familiar frigid tone crept into his voice. 'In such circumstances failure is not admissible.'

'Do we know the commandos' objective?' Although he was not as afraid for his own skin as Stolypin supposed, Makarov saw no reason why the directive should be vague. From the day he joined the Army he had been accustomed to receiving orders on every detail. It was the Red Army's way of life, as well as the Party's.

'The Isfjord radar site remains the obvious objective. They could be taking an inland route.' Not only did Stolypin dislike being wrong, his supposition was justifiable. The whole crisis was now centred there. 'However, they could just as well be an observation party, reporting on our dispositions. What you have to do, Comrade Colonel, is track them down. Logically, they must approach one or other of the settlements.'

This was as much official help as he would get. When Makarov left the office, instead of being taken straight to the airfield, he had himself driven the short distance down to the harbour beneath the overhead coal conveyor buckets past long warehouses painted pastel colours, all dulled with grime. Longyearbyen had coal dust everywhere, mixed into the road surface, obscuring windows, blackening residual mounds of snow. Even the sparse tufts of grass, still brown from their long winter burial, were stained. It was an extraordinary contrast to the pure white of the glacier, just visible beneath the cloud at the head of the valley.

The harbour itself was little more than a broad wooden jetty in deep enough water for cargo ships to come alongside. Tied up on an inner mooring was the *Northern Light*, two sentries conspicuously guarding her. They came to attention as Makarov approached, their greatcoats swirling. Even though it was still bright daylight, the temperature fell at night.

Makarov went aboard, his boots clumping along the short gangplank. On Stolypin's orders, the *Northern Light* was being kept exactly as she had been when captured. The Norwegian flag still flew limply from her mast. As far as the locals were concerned, she had been arrested for an unspecified offence.

Weston, asleep in the cabin, sat up as Makarov entered.

'What do you want?' he asked in Norwegian. If the Russians were prepared to keep up the pretence, so was he.

'Show me the hold.'

'Not again!' Reluctantly Weston got up, pulled on his trousers and led the way down through the engine room. He recognized Makarov's badges of rank, and when the Colonel had finished examining the wooden structure of the bunks, Weston thought it worth trying a question.

'How long are you keeping us here?'

Makarov laughed. 'Until your navy arrives, perhaps.'

It was a double-edged answer and Weston knew it. Whose navy? Norway's?

'In return, may I ask you something?' Makarov's attitude was politer than an interrogator's would have been. He knew precisely who Weston was, his rank, his career in the Marines, where he had sailed from originally, who owned the ship; and he respected the daring of the enterprise. It had been foolhardy, but brave.

'You can try, Colonel.' Possibly the question would explain what the Russians were playing at. Weston was still amazed that they had not sunk the *Northern Light* and imprisoned him.

'You have accommodation here for fifteen or sixteen men.' Makarov watched Weston's face, looking for confirmation of his estimate. 'How many did you put ashore?'

'We were fishing.'

'But got caught yourself, eh?' Makarov toyed with the idea of spinning out a friendly conversation in the hope of gaining a few clues, then dismissed it. A commando with Weston's record would let little or nothing slip. 'Never mind, Major,' he said genially. 'It could happen to any of us.' By God, that was the truth, too.

However, as the helicopter clattered its way back to Heerodden later, he knew the visit to the ship had been worthwhile. There was sleeping accommodation on the *Northern Light* for sixteen but not enough room for the gear they would need in the mountains. He guessed they could not have numbered more than ten.

That evening Makarov retired early. Whoever had furnished the room had provided only thin curtains and whereas Arctic veterans grew accustomed to the continuous daylight, he was kept awake by it. His thoughts wandered inconsequentially, speculating on when the seagulls slept, how much heat escaped through a reindeer's uniquely insulated coat, what route he would take himself if he were attacking the Plata Berget. Eventually he draped a spare blanket over the curtain rail, determined to rest properly before the next photographic reconnaissance report arrived. He was unlikely to get much sleep for the next two or three days and age was beginning to blunt his old resilience.

For the second night the team were dug into snow holes, this time against the side of the mountain bastion separating them from the Plata Berget. Tomorrow, Friday, would be the toughest stage of the long approach march. Accordingly Peterson decided to use these few hours of protection from both the cold and possible enemy patrols to give each man a last briefing on his role, just in case there were unforeseen problems at the final laying-up point.

He began, naturally, with his second in command, Howard Smith, who shared a hole with Trevinski. Peterson squirmed in and then sat on one sleeping shelf, while they faced him from the other. The candle flickered as he unfolded his map, casting grotesque shadows on the opaque snow walls, and he used a torch to supplement its light.

'I'm going to have Captain Mydland both trailbreak and navigate tomorrow,' he explained, tracing the only possible route up the crazily steep mountain to the horseshoe-shaped glacier above Longyearbyen, across one part of it and then skirting the moraine of the other arm to reach the radar site. 'Don't want to find ourselves coming down towards the township by mistake, right?'

'Damn right, Colonel,' Smith agreed.

With his goggles and parka hood off, Smith's face was fully revealed and Peterson noticed that his eyes were slightly screwed up with the kind of quizzical look a footballer might give a team coach who was changing the tactics in the middle of a match. Trevinski, on the other hand, was hunched forward impassively, all attention.

'Now, when we come to the attack,' Peterson went on, 'I'm going to divide us into two groups, the same way we trained. I'll have Burckhardt, Johnson, Millar, Neilson and Captain Mydland with me. You and Trevinski take the Dragon, give fire support and fix the demolitions.'

Smith said nothing, his expression hardening. Peterson took no heed. If Smith didn't like the way this assault was going to be carried out, that was too bad.

'I don't reckon there's going to be many gooks on that mountain, if any at all. The satellite pictures showed no defensive digging.'

'We can't be sure, sir, can we?' Trevinski said.

'No, we can't. But whoever's there, I want to use this poor visibility to get in so close we can overwhelm them and they have no time to cry out. Then we can fix the charges nice and easy, fall back and call the chopper before the gooks down below know what's happening.'

'You mean, Colonel,' Smith said flatly, visibly holding himself in, 'we aren't going to use the Dragon to take out the radar?'

'Not if we can avoid it.' Peterson tried to keep his own voice sympathetic yet firm. He knew an outburst was coming.

'Goddamn it, sir,' Smith exploded. 'You mean we hump this fucking launcher and these rounds two and a half days and we're not even going to fire them? Holy shit, the goddamn plan was take advantage of the missile's range. That's what we planned for.'

'You did, Captain. Not me.' Peterson glared back at his subordinate, now equally angry at this officer's intemperate language. 'Be clear, Smith, we have this foul weather, we're going to exploit it. We're going to take out this radar in such a way those Soviets down on the airfield think maybe it's blown a fuse or something. Until they find it's one hundred per cent destroyed. By then we'll have gone. This long-range attack's strictly for if the weather clears and we've no other option.' His own voice was clipped and unyielding. 'Could also be useful to give covering fire if there's any problem with our getting lifted out, right? That's my plan and let's work on it.' The put-down was as close to an outright rebuke he could give an officer in the presence of an enlisted man, even in as close-knit a unit as this.

Smith bit back a retort. He had long cherished a vision of himself and Trevinski skiing silently into position under cover of the mist and waiting for a patch to clear, the launcher resting on its bipod and his shoulder. Trevinski would load, as they had so often practised, he'd take aim through the optical sight, and then would come the spine-chilling, roaring whoosh as the Dragon fired and the missile streaked across the snow to erupt in a great glorious blossom of flame as it hit the radar. That one explosion would make the cold and fatigue and the months of training all worthwhile. They'd withdraw into the mist and vanish. But now this goddamn, scar-faced ex-sergeant had fouled it up. Who the hell did Peterson think he was anyway? A guy who only joined the Marine Corps to avoid a jail term.

Peterson saw the disappointment and antagonism in Smith's face and guessed much of what he was thinking. His fists clenched involuntarily. Ever since that day when, as a seventeen-year-old, he'd beaten up the layabout pestering his mother, and the judge had preferred the layabout's evidence to his, he'd been working out his own adjustment to society. The Marine Corps had done most of the work for him, Nancy was completing the job. But he still instinctively reacted violently to the kind of ride Smith was giving him.

'Maybe,' Trevinski said cheerfully, 'we should leave the John Wayne stuff to the Green Berets, heh? Me, I'd rather get back to a hot broad any day.'

The tension broke. They all grinned. 'You're the expert, Corporal,' Peterson said gratefully. Trevinski's womanizing was legendary.

'Wouldn't know about that kind of thing, would you, sir? Being a married man.'

'No, sir.' Peterson shifted, delighted at Trevinski's good sense. If nothing else came out of Virginia Ridge, the Pole would emerge a sergeant. He turned to Smith. 'We'll saddle up at five, Howard.' He used the Christian name deliberately, to demonstrate that, so far as he was concerned, the clash of personalities was over. 'Now I'd better brief the others.'

'As you wish, Colonel.' Smith's response was unyielding, its implication clear: have things your own damn way, I wash my hands of the outcome.

Peterson crawled out, restraining a fresh surge of anger,

and made his way to the snow hole occupied by Millar and the Norwegian. The latter's key role would be over once he had led the way to the Plata Berget, as was the Briton's already, but with so few men both had to be allocated new tasks. Using a sketch of the site, Peterson explained how he wanted them to tackle the hut, while he and the other three seized the radar and Smith and Trevinski covered them.

'We'll give them a couple of grenades as a wakey-wakey, sir,' Millar remarked happily. 'Bet a pound to a penny they'll be sound asleep and won't even have locked the door.'

'Fine.' Peterson discussed the basic tactics of the approach for a few minutes, then made his way on to Johnson, the medic, and Neilson, the linguist who doubled as an explosives expert. The radar truck was almost certain to be unmanned and he wanted to waste no time placing charges on it which they could wait to detonate until the helicopter was on its way to extract them.

'Let's be clear about this,' he explained to the two men, 'what I want is for us to be lifted out before the Soviets know what's happened. That way not only do we save our skins, whatever the President tells Moscow on the hot line will have maximum effect. There's one hell of a lot riding on this raid.'

Finally, he returned to his own snow hole, hoping that by now Burckhardt might have established communications with Bodo. To his concern, the radioman had not. There was still appalling interference.

'Well,' he said, forcing optimism, 'let's catch some sleep. The next call isn't due till we reach the LUP anyway.'

'Shouldn't last much longer, sir,' Burckhardt said, equally attempting to be cheerful. 'Conditions must be changing soon.'

Sure, Peterson thought as he wriggled into his sleeping bag, but what we need is both the communications and the cloud.

*

Every night and every morning, before she went to bed and at breakfast, Annie listened to the Oslo radio, in case the news bulletin contained a message.

For the past three days the reception had been appalling. Tonight it was better and the news itself momentous.

'The Norwegian, American and British governments announced today that a naval blockade will be imposed on Svalbard effective from midnight GMT, that is, from zero one hundred local time. The Justice Ministry appeals to all citizens to remain calm. The Ministry asks us to repeat this advice. All citizens should remain calm. According to official sources the blockade is a response to the Soviet Union's refusal to recog – ' A loud and persistent buzzing obliterated the broadcaster.

Annie realized angrily that the Russians must be jamming the wavelength, but that was unimportant; any repeated notice from the Justice Ministry was a message to her. Leaving the radio on, as a normal person might, she fetched the special set from its hiding place, rigged the aerial and tapped in the Morse code signal to indicate she was listening.

Two minutes later the encoded reply showed as a flashing light. It took her more than twenty minutes to unscramble, first reading the Morse code, which was simple, then laboriously using a one-time pad to convert each letter until short sentences emerged. She read the result with a chill of apprehension.

'Activate Wolfhound' – that meant Lars-Erik, the football player and miner – 'to organize civil disturbance Friday and Saturday. Top priority keep Russian police occupied next 48 hours.'

The order scarcely needed memorizing. She burned the sheet of paper in an ashtray, crumpling the twisted black fragments until they disintegrated and her fingers were black too. Then she washed her hands, swilled the ash down the lavatory, and telephoned Lars-Erik. It was a risk, since the only telephone in his bachelor quarters was a communal one, but less conspicuous than going out herself late at night.

'I'll do my best,' he assured her, excited to hear about the blockade. 'Those bastards who killed Sven deserve everything that's coming to them.'

Despite his aggressive confidence, Annie was trembling when she put the phone down. In the West a civil disturbance usually ended with nothing worse than a few broken heads. In Russia life was held cheaper. Whatever made a riot necessary, she prayed it was worth the danger.

Makarov woke early, theories about the raid circling in his mind as relentlessly as hawks in an African sky. He lay staring at the ceiling, trying to think himself into the enemy's position. Suppose this was simply an observation party, what would they report on? Surely only those installations around Longyearbyen from which the locals were kept away, in other words, the airfield? Or the radar? No, that would be pointless. American satellite photography would have identified its detail long ago. But, as he had always thought, they could be aiming to blow up the radar. Possibly in conjunction with a blockade to prevent installation of the Isfjord masts. God in heaven, that made sense.

When, over a hasty breakfast, he learned that the Norwegian government had set a starting time for the blockade of midnight GMT and the US Fleet would back the Norwegian Navy, he knew he must be right. It was to be a strictly limited warning action. But where the hell was the raiding party? He hurried to read both the latest photographic interpretations from Archangel and the weather forecasts, though he had only to look out of the windows to see the improvement.

During the night several signals had come via the satellite communications system from the Air Defence Command. The final summary was a masterpiece of evasiveness, acknowledging no responsibility for faulty interpretation and blandly stating that a series of photographs were required before proper analysis could be made. Immediate investigation was now required of several heat sources. One group in particular had been travelling north-east through the mountains but had completely vanished from the last infra-red pictures. Neither the speed of movement nor the disappearance was consistent with the behaviour of animals. The trace could be of a patrol which had laid up on the way.

Makarov fixed the various positions on the map, his excitement mounting. The track led from near the coast towards the Colesdalen. It was a way he might have taken himself to get above Longyearbyen unobserved. He surveyed the markings on his maps again, checked the times of the reports shown against them and came to a decision: he would go straight to the Colesdalen and investigate for himself, leaving the Spetsnaz company commander in charge at Barentsburg. At least he had traversed the mountains behind Longyearbyen before, which was more than the Spetsnaz had done.

Twenty minutes later the big helicopter was swaying into the air. A twenty-man fighting patrol was squeezed into the cabin, their equipment piled on the floor between them. Makarov was up with the pilots, peering out as the machine tilted forward and gathered speed. It raced along the shoreline at 2 kilometres a minute, banking as it turned right into the Colesdalen valley, past the semi-derelict Colesbukta jetty and the abandoned mine workings. A handful of border police gazed up and waved as they flew overhead. The low mist had completely dispersed and the helicopter rocked and shook in the wind channelling down between the mountains. Ahead, the high snow slopes were one with the underside of the cloud, a consolidated white oblivion. Heavy snow was falling up there.

The pilot glanced at his neatly folded map. 'Five kilometres to run, Comrade Colonel.' He swivelled the pitch control and the helicopter slowed, its tail dipping slightly, the co-pilot scrutinizing the snow beneath. He and

Makarov saw the trail at the same moment and both shouted. The pilot brought the machine to the hover, high enough up for the downwash of the rotors not to disturb the snow. The trail was unmistakable; two deep indentations like single tracks, wider than one man would have made, while on both sides were the round holes made by ski poles. The trail led in a gentle curve across the width of the valley, its lines broader and more ragged as it reached the upward gradient. In places its makers had left a herringbone pattern where they ascended a hump. There was no question about their direction.

'Follow it up,' Makarov ordered. Every kilometre the helicopter could fly would save the patrol significant time.

Snow began to blur the windscreen, whipped by the slipstream. Half a minute later visibility was down to 50 metres, the way ahead completely whited out. The pilot descended to a few metres, then was forced to hover.

'There's not enough horizon, Colonel. Do you want to land?'

Makarov looked at the trail vanishing into the blizzard and knew that the decision he made now would be both crucial and irrevocable. If he dropped the fighting patrol here, they would be lost to him until the storm was over. They could not be redeployed in the light of new infra-red intelligence. But in two or three hours the tracks themselves would be buried. He hesitated only briefly, as the helicopter rocked in the turbulence. He was going to win this battle, damn it, and the devil chafing at his back, demanding success, was doing so not for the Party, much less for Stolypin, but for his own self-respect.

'Land,' he commanded, and retreated to the cabin to shout last instructions to the officer. 'Follow the trail. If you sight the enemy, maintain contact and radio for instructions. We want prisoners, live prisoners. Understand?' He had to half scream as the doors were opened and a great rush of noise and freezing air swept in. 'I'm going to the

Plata Berget. Their objective must be the radar. If we lose touch, force them out on the plateau. They'll be pinned down between us.'

The men were jumping out, weapons in hand, stumbling in the snow as they ran clear of the rotors.

'Good luck!' Makarov slapped the lieutenant on the shoulder. 'Now, get going.'

The pilot lifted off, creating a blizzard of his own, and Makarov leaned out of the gaping doorway, his thumb raised. The white-clad officer glanced up and saluted. As the helicopter whined away, the men were fixing on skis and adjusting packs. They were the élite of the Red Army. If they couldn't do the job, no one could.

On Makarov's return to Barentsburg he was asked to contact the General's office immediately.

'An order from the Comrade General,' said the aide who answered. 'We have trouble here. You are to send a detachment of Spetsnaz to assist the border police in defence of the airfield.'

'What's going on?' Makarov demanded. 'I'm short enough of men as it is.'

'The miners are rioting,' the aide snapped and rang off. He could afford to be abrupt, knowing Stolypin's opinion of the Colonel.

This information clinched Makarov's belief that a climax was close. He despatched a platoon of Spetsnaz as directed, though he was tempted to argue the toss. Out of his 140 Spetsnaz, thirty-odd were deployed at the Isfjord radio site and a similar number at the helicopter base. The fighting patrol was irretrievably committed. Now he was losing a further thirty. He was left with the few soldiers in the company headquarters and the patrol detachment of the border police troops. He had enough men to defend the Plata Berget, but only just. All other options were lost. He would have to stake out his pre-planned killing ground on the mountain and trust that his intuition was correct.

*

The breakfast-time broadcast was obliterated by jamming. Annie decided there was unlikely to be another message for her and the fewer transmissions she made the better. She pulled on her coat and left the apartment. If there was a riot, she might need to protect the bank rather than open it, though it was merely a portable, prefabricated office, like the ones on construction sites.

The housing block was only a few yards from the main road, which led up the side of the valley to the shop, her bank, and the miners' main mess hall, the Store Messe. The road was free of snow, though the banks of snow cleared earlier in the year were ranged along the sides, dirtied by coal dust and grit. To her surprise, groups of miners were already gathered at the parking space by the shop, some standing on the piled snow, others milling around, all visibly waiting for some kind of collective decision.

'Why aren't you at work?' she asked one man in a red and blue striped anorak, who was stamping his feet against the cold.

'Haven't you heard?' He looked at her with curiosity. 'We're being blockaded, I mean the Russians are, and that bastard general of theirs has had the neck to stick a notice up at the Store Messe saying if there are food shortages from now on, we've Oslo and the Americans to thank. Bloody nerve!'

'Well!' She pretended ignorance, thinking that Stolypin could not have given Lars-Erik a better lever.

'They've no right to be holding that fishing boat either,' cut in another miner aggressively, his hands deep in his pockets. 'She's Norwegian and registered in Hammerfest, isn't she?'

Annie made no comment. The crew were known to be under arrest on board and she suspected they might have been on a spying mission.

'We're not standing for it,' the first miner said. 'The stewards are meeting now in the Store Messe, deciding what to do.' He eyed Annie's long brown hair and perky fur hat.

'You're the bank girl, aren't you? Better go and find out what they want from you. If the stewards call for a stoppage, we'll want one hundred per cent support. Everything will stop, power and light included. That's my guess.'

She hurried along to the large, drably painted building, trotted up the steps and through the small foyer into the vast dining room with its rows of orange-coloured tables and long self-service counters on the right. At the far end, a group of miners was gathered round the local union leaders, standing on a platform improvised from tables. Behind them on the wall hung a portrait photograph of the King in admiral's uniform and alongside it a picture of Lenin, which had caused much resentment when the Russians put it there.

The impromptu meeting was ending. The senior union steward raised both hands in front of him.

'We're agreed then, brothers. A total walkout and a protest march on the Governor's office. Let's go.' He turned round slightly awkwardly on the table top, reached for the photograph of Lenin and turned it face to the wall. A shout of applause and clapping greeted the gesture. Then he jumped down and led the way confidently out.

As Annie stood aside to let them pass, Lars-Erik came up to her briefly. 'Their blood's up, all right. Hadn't you best stay at home? There's going to be trouble.' Though he spoke quietly, his voice sounded as though he was happily anticipating a fight.

One of the women who served in the shop joined them. 'Come on, Annie, you're not going to miss this, are you?'

For a few seconds she was torn between the possible need to look after the bank and her obvious duty to report to Oslo later about what happened to the demonstration. The shop assistant answered her unspoken query.

'No one's going to raid the bank,' she said cheerfully, 'not today. Or if they do they'll regret it.'

'See you later,' Lars-Erik said, and left them.

Outside a procession had begun down the long road towards the centre of the settlement. Annie and the other woman followed. As they passed on down the hill more men joined in until the crowd was a surging phalanx. Someone began singing the national anthem.

As they crossed the valley, a small helicopter appeared overhead, circling noisily. The marchers at the head of the ever lengthening column stopped, and across the bend in the road Annie could see the few remaining Norwegian police in their blue uniforms talking to the leaders. Then the police stood aside and the march continued, the chanting and singing growing louder and more determined.

When the crowd slowed down again, outside the Sysselmann's office, she was perhaps 300 metres behind. The church, strikingly modern, was only a short distance away, and she had positioned herself there to have a reasonable view. She saw a large number of Russian border police with rifles in their hands, ringing the offices. The leaders of the march halted and the crowd began to fan out across some waste ground as those at the back pushed forward.

There was a lot more shouting. Annie could not see anyone being allowed up to the long wooden office building. The crowd went on pushing forward. Suddenly a single shot rang out, sharp and frightening. There was a second's silence. Then a great roar of fury from the miners was followed by a fusillade. She heard bullets crack through the air overhead. The crowd recoiled. Several men ran off up the hillside and began throwing chunks of rock down at the Russians. She saw one stagger and fall. A hand clamped her shoulder.

'Come with me,' a man's voice said. 'There are things to do.'

It was impossible to see the township from where the *Northern Light* was moored on the inner side of Longyearbyen's wooden quay. Sheds and warehouses obstructed the view. But the distant shouting of the crowd was audible and aroused the curiosity of the two Russian police guarding the ship. They went forward, confident that their three captives could not escape, and gazed up the valley.

Inside the cabin Weston rose from his seat, opened the door gently and listened.

'There's something going on,' he told Clifford, who was resting his injured leg on the leathercloth bench. 'Sounds like a riot.'

'*Niet!*' One of the guards swung round, gesturing with his rifle. Weston retreated.

Moments later another policeman came running down the quay, boots pounding. He stopped, out of breath, and called to the guards, just as a burst of firing echoed from the town. A brief and excited conversation ensued. Then, watching through the window, Weston saw one guard jump across to the quayside, look back quickly at the cabin as though reassuring himself, and double away with the other man.

'They've got real trouble, Craig. This could be our chance.'

They had been discussing ways to escape whenever they were alone, certain that Peterson and his team faced disaster if he was not warned. The *Northern Light*'s normal radio remained intact. If they could only get free for ten minutes they could contact the Norwegian Navy.

'If we could clobber that joker,' Weston went on, 'by luring him in here . . . you could feign sick, Craig.'

'Sure.' Clifford began to rearrange his position to make it seem he had collapsed.

'What do you want from me?' Paulsen asked.

'Pretend to be helping him.'

From the town came the unmistakable sound of a shot.

'Starter's orders,' Weston said tensely, swung the door open noisily and shouted, 'Comrade, come here!'

The guard looked round reluctantly.

'Comrade,' Weston repeated, utilizing the half dozen words of Russian he knew. 'Here. Quick.'

As the guard came back along the narrow companionway

past the wheelhouse Weston stood aside, as if from courtesy, and pointed through the doorway. God, if the man fell for this, he was a fool. But Clifford certainly looked ill. Paulsen was bending over him attentively.

While the Russian stood eyeing Clifford with evident suspicion they all heard the volleys of shots from the town. Distracted, the guard moved towards the window. As Paulsen seized his chance and pinioned the man's arms, Weston lifted a wrench they had been using earlier and slugged the back of the guard's head. His fur hat absorbed part of the blow, but not enough. He collapsed backwards onto the table.

Weston surveyed the quayside. It was deserted. 'Come on,' he urged. 'Strip the bastard.'

Five minutes later Clifford, dressed in the Border Police Troops uniform, was thrusting a rifle into Paulsen's stomach, apparently ordering him to cast off the ropes. Weston started the engines and they went slowly astern, then turned, the water churning behind, and set off at the *Northern Light*'s maximum speed of 11 knots down the Isfjord towards the open sea.

9

The higher they climbed, inching their way up the steep boulder-strewn snow slope, the more savage the wind became. The only consolation, Mydland thought as he led the way, was that it seemed to have shifted in direction and was coming from behind, perhaps funnelled that way by the peaks. What would happen at the top was anyone's guess. There the geography changed. They would cross the ridge onto the 4-kilometre-long glacier leading down to Longyearbyen: a different valley, maybe a different wind. But this was bad enough. It was a gale, tugging at his clothing and the hood of his parka, beneath which he pulled a woollen scarf up over his mouth and nose. After every few steps he halted, turning his goggled face into the blast, to glance at the shadow figures following, and the ice spicules stabbed at any exposed flesh like needles. The visibility was down to 5 metres.

Peterson, third in the ragged line, shrilled his whistle for a halt and laboured his way up past Burckhardt. Even moving out of the direct trail and back was a costly additional effort. This was a south-facing slope and under the fresh snow there was old ice, melted by the sun at midday and refrozen. In places the wind had whipped it bare again and he slipped on a patch now, hastily edging his skis to stop himself sliding, recovering his balance just in time. He reached Mydland, stood as close as he could and yelled.

'How far d'you reckon?'

Most of the sentence was stolen by the gale. All Mydland heard was 'far'. It would be futile trying to get out a map. He fumbled for the altimeter, realizing how slow his movements were. With this wind the effective temperature must be down around −28 degrees. Frostbite was becoming a serious risk. In such cold even the most experienced men had to expend most of their energy on simple survival. The altimeter, cupped in his hand for protection, showed 1900 feet. They had at least another 250 to climb, or farther if in his struggle to follow a compass course he had strayed from the direct route to the saddle between the mountains.

'We should stop,' he shouted, pointing upwards.

Peterson followed his gaze, shielding his goggles with one hand. A ragged line of rock showed dark above them through the driving snow. They had reached another of the strata they had seen from across the Colesdalen. The last 500 feet had taken two and a quarter hours because they had twice been obstructed by similar outcrops of rock, which crumbled dangerously. His toes were already numb and he found it hard to think clearly. To stop was against all his instincts. But if he was losing feeling, so might the others be. Mydland did not seem the quitting kind either. Reluctantly he signalled agreement.

For an hour they huddled in their individual tent sheets, utilizing such meagre shelter as the rock provided and dug as deep into the snow beside it as they could get. Their packs and equipment also helped break the force of the wind. Squatting there, trying to keep warm, Peterson realized that he had definitely lost all feeling in his big toes. He ought to take his boots off to let the blood circulate, bringing on that agonizing pain which would spell recovery. But it was out of the question. He tried to wriggle his toes inside his thick socks. Still no feeling. Better to lose your toes than to lose your life, he told himself. When they reached the LUP he would get the damn boots off. We're going to make it fine, he reassured himself. It wasn't ten in the morning yet, this storm must blow itself out and at the worst there were only another seven clicks to go. The important thing was not to let the team's morale slip. 'This is our travelling road show,' his old company commander used to say, 'and we're going to keep it on the road.' The lesson to be taken was the same he had learned as a newby in the 'Nam all those years ago. Willpower was what got you through the bad humps. Except that when you were in command you needed a larger ration, so you could share it around.

Being a successful commander was an art that could not be wholly learned. There had to be a great spark inside yourself, not just organizational ability, not just a never ending readiness to lay your own reputation on the line, but something more. Men had to be willing to risk their own lives on your decisions: as these men were now about to do on his. Peterson pondered Smith's reaction to his briefing in the snow hole last night. In no way had he carried Smith with him. Yet he could not believe this was due to any failing of his own. Smith's character had a sullen, angry, selfish streak which had first revealed itself up in the High Sierras. It had surfaced again in Scotland, when he fired the missile ahead of Clifford's order because the mist was clearing. Jesus, Peterson thought, if Smith had done that in a real fighting situation the team would have been slaughtered. He had not overdone the reprimand on that occasion. As for last night, the man deserved to be reduced to the ranks for the way he'd said 'As you wish, sir.'

Yet, Peterson reflected, am I right? Smith's determination had made him a football star. When it matured he might emerge a successful general. Military history described Patton as a show-off with his pearl-handled revolvers, while MacArthur had taken self-opinionated action to the point where President Truman sacked him. None the less, generals George S. Patton and Douglas MacArthur had been among the greatest commanders of the century. So where the hell did that place Howard Smith? With four stars burning a hole in his pocket, or consumed by a desire for personal glory which might wreck this whole damn operation if it wasn't checked?

Wrapping the tent sheet tighter around him, Peterson tried to disregard the numbness of his feet and killed time chewing over relationships. These weren't his strong suit either. He wished he had discussed Smith further with Nancy that evening in London, but he had been fearful of giving away anything about Operation Virginia Ridge. A wife, however loyal, had no 'need to know'. Hell, no wonder either that she wanted him out of the Marine Corps. That unquestionably was a problem waiting for when all this was over. He had baulked at considering it over the past weeks, deliberately unwilling to concentrate on anything except this mission.

Now, in the isolation of the mountainside, the wind whirling snow around him, he suddenly found himself willing to think about other careers. What careers? Retired generals could become vice-presidents of big corporations, helicopter pilots or engineer officers had technical qualifications, even an Air Force sergeant probably had better civilian capabilities than a recently promoted Marine Corps lieutenant colonel whose basic trade was killing people, or organizing the same. He was at a halfway house, with no master's degree in management and not enough seniority to be able to jump straight to a company board. Secretary of a golf club, maybe? Selling insurance round bases like Fort Bragg, using old military contacts? Jesus Christ, no! This was his chosen career, squatting here in a goddamn freezing wilderness, but at least carrying the can for something worthwhile. He realized he had made up his mind. Nancy was going to have to live with the Marine Corps either until he was senior enough to opt out gracefully or at least until he had earned a larger slice of pension than he'd be entitled to right now and could pick and choose in the world outside.

Achieving this decision lightened Peterson's mood more than he would have expected. He turned his thoughts back to the next hump, wiped his dark goggles with his glove and peered out into the storm. It was definitely abating. Before long they would be able to saddle up again.

'Come with me,' the man had said, gripping her shoulder. 'There are things to do.'

Annie had instinctively wrenched herself free and swung round to find it was the shop manager.

'The hospital won't be able to cope. We must get emergency dressings organized.'

He had hurried her back up the road to where his car was parked and they had driven to the shop and the mess halls, collecting first-aid boxes and medicines, then set themselves up in the House. As a social centre it had a foyer with its own shop and a small post office. This was where they prepared to receive casualties.

As the manager had foreseen, the hospital was only able to take the seriously wounded. Soon other marchers began to come in, supporting their injured friends. No one knew how many had been killed, only that the border police had shot into the crowd, deliberately and repeatedly, as well as over their heads. They were all grimly revengeful.

As she was helping carry a man in, Annie heard the familiar noise of helicopters overhead and looked up. Two big, camouflage-painted machines were flying up the valley together. Hanging below inside cargo nets were snow scooters. She counted four, all brightly painted and presumably confiscated.

'Bastards!' shouted the man with her.

The helicopters disappeared up the valley, where thick cloud hung over the glacier. She did not worry about their mission until twenty minutes later the same two came back down without the snow scooters. Initially she was perplexed. Why should they be taking troops up into the mountains when the riot had been in the settlement? Then she guessed that Paul Mydland, or someone else from the *ski-*

jeger, might be up there again. The thought shocked her. She made an excuse to go back to her apartment, took the radio from its concealment and began laboriously encoding a warning. After fifteen minutes no reply had come. She sent the message again. This time it was acknowledged, though nothing more. She hurried back to assist with the injured. It was the only useful thing to do.

Towards the end of the morning Lars-Erik came to the House and spoke to her unobtrusively.

'The second shift are refusing to work,' he said. 'They've plans to barricade off part of the town. You'd better stay indoors. There could be worse trouble later.'

General Stolypin was not the person to panic when things went wrong. The border police lieutenant standing woodenly at attention in front of him in the Sysselmann's office evidently was. The young officer's stolid expression was failing to mask a far worse fear than had beset him earlier in the morning. His tight-set lips trembled perceptibly.

'You opened fire into the crowd?' Stolypin repeated, genuinely incredulous for once. 'You gave the order to fire at them, not above their heads?'

'Those miners were out of control, Comrade General.' The lieutenant could not keep his voice steady. 'We did fire over their heads at first, but they didn't halt.'

'That does not tally with other reports I have received. Do you realize you have endangered your country's international reputation; you have put at risk the entire Svalbard project? Your orders were to turn back the demonstrators, not massacre them.'

The officer stood as rigid as a doll, his eyes fixed on the window behind, not daring to look his superior in the face. Stolypin would have liked to walk round the desk and rip off the idiot's insignia of rank, to degrade him here and now.

'You will be court-martialled,' he said icily, 'and are under open arrest. Dismiss.'

'Fool,' Stolypin thought as the lieutenant saluted, turned and marched out. 'Double-dyed idiot.' The best to be hoped was that the man would utilize his open arrest to commit suicide. At least that would confirm his guilt instantly. Meanwhile, despite the humiliation involved, the mine manager would have to be informed that a Russian officer was being held responsible for the deaths. But there was little chance of that announcement defusing the situation. An iron grip would have to be kept on Longyearbyen.

Stolypin turned his attention to the other serious problem the morning had brought: the escape of the *Northern Light*. He rose and studied the same map on the wall to which Makarov always referred, slowly recovering his temper as he read its message. The Isfjord was shaped like a hand with the fingers of subsidiary fjords outstretched. The wrist, where it joined the sea, was narrow, a mere 15 kilometres wide. He knew it was still partly obstructed by floating ice. The *Northern Light* could be intercepted any time he wanted.

But did he want her intercepted yet? No, he reflected, emphatically not. He could be 100 per cent certain that the British commando major would make contact with the blockading ships and warn them that the operation, whatever it was, had been blown. There should be a fair chance that the raiding party would be recalled immediately and attempt to board the trawler, be picked up by submarine or even airlifted out. The moment that was arranged, further radio contacts would be inescapable and the electronic intelligence would pinpoint the raiders' location in seconds. For the first time today Stolypin smiled, a hard-edged, ironic, private smile. There was still half a day left before the NATO blockade began: twelve hours in which to capture those damn commandos. He would let the *Northern Light* go her own sweet way and help him to the best of her ability.

None the less, once he had called his aide and given the

necessary orders, Stolypin felt his good humour die. He was running out of time, short of men, and too many of the Spetsnaz were irretrievably committed in the mountains. God in heaven! How had he allowed a situation to develop where, unless the enemy saved his bacon for him, everything depended on Makarov trapping the raiders? Makarov had been set up to be the scapegoat, but the riot had altered all that. Now he could only be a saviour. Indeed, he had to be. It was impossible to prevent news of the shooting leaking to the outside world. Too many local people sought solace from the winter dark by operating ham radios. The outcry would be universal and if the capture of the raiders was not achieved by midnight, he felt sure the Politburo would lose its nerve over the blockade. If that happened, they were all destined for the scrapheap.

Stolypin shuddered in spite of himself as he considered his future. True, his career would not be as brutally terminated as in Stalin's days when failures were liquidated; the disgrace would be more subtle. But he had no desire to be banished to some remote region, nominally in charge of a run-down minor industrial plant which never would, and never could, meet its production norms, where further failure was predictable and intended. In the Soviet hierarchy the descent from power could be slippery and fast. He clenched his fists in frustration, seeing no way in which he could influence the next twelve hours' events unless this cursed weather lifted.

Soon after midday the wind began to drop and snow fell less thickly, improving the visibility to 40 or 50 metres. Peterson ordered a redistribution of the missile components and the team shook themselves, flexed half-frozen muscles, cursed routinely, and set off again. Even Howard Smith had given up any attempt at shaving in these conditions. The rest were grateful for what protection two days' stubble gave their faces.

They reached the crest 200 feet higher up than planned. The glacier stretched away, what they could see of it, in a smooth expanse of snow. Except that between its edge and them there yawned a deep crevasse.

'The summer melting must have begun farther down,' Mydland explained. Glaciers were in constant movement, calving huge chunks of ice into the fjords in summer, though this one never reached the water.

'Sure, but how the fuck do we jump this bastard freight train?' Trevinski demanded. 'Wait for next winter?'

'Take it easy.' Peterson intervened. 'We'll work our way down a little.' Trevinski might be the team's humorist, but he didn't want his remarks getting under the Norwegian's skin.

Eventually they found a place where the crevice between the rock face and the snow-laden ice was only 4 feet or so wide. Smith removed his skis and leaped it, slipping on the far side, but successfully scrambling up again. He drove a spike into the ice and secured a safety line. Then Peterson's sense of caution prevailed.

'Lash pairs of skis together,' he ordered. 'Make a bridge.'

Fixed like this, the skis constituted two narrow parallel planks. One by one the men edged their way across on hands and knees, the skis flexing beneath their weight.

The last man to go was Johnson, the medical corpsman, his heavy rucksack giving him the appearance of a grotesque hunchback. His hands were touching the glacier when one of the ski lashings slipped, not much, but enough. The ski shifted sideways, he lost his balance, grabbed for a handhold and fell, the safety line round his waist digging tautly into the snow as the others strained to hold it.

'Stay still,' Peterson shouted. 'Push yourself away from the sides as we pull.' He was frightened the rope could fray or another man be jerked down after Johnson. Hauling on the rope, sweating, they rescued him.

'Sorry, fellers,' he muttered, examining himself on the ground. He was bruised and cut.

'Could happen to anyone,' Peterson said. 'Let's get going again.'

But Johnson only pushed himself upright clumsily. 'Colonel,' he said with embarrassment, 'I guess I hurt my wrist there. Could someone fix a bandage for me?' He held out his left arm, grinning painfully. 'Lucky it's not the right hand, eh? The box is in my pack.'

Millar pulled off his gloves despite the cold, felt the bones and knew one was broken, then made as quick a job of strapping the wrist as he could, helped by Burckhardt. 'There you are, mate,' he said cheerfully, blowing on his fingers to warm them again. 'All part of the service.'

'Thank you, Sergeant.' Peterson meant it. Millar was brisk and efficient. He hoped to God this would be the last accident, though it could have been one hell of a lot worse. Johnson could still ski and work a radio. Burckhardt would act as medic.

As they trekked across the glacier, a gratifyingly easy run slightly downhill though the wind was now against them, the snowstorm began to lessen and when they reached the far side the crevasse was only a 2-foot-wide crack. With Mydland still breaking the trail, they worked along the side of the ridge, maintaining complete silence and moving with caution. Longyearbyen lay hidden 1800 feet below. They were now maybe 6 kilometres from the radar and 4 from the LUP. It was after 1600 hours and operation Virginia Ridge was near its climax.

The radar was in a camouflage-painted van on four wheels, with a door in the side and two large elliptically shaped dish aerials. One, mounted on the front end of the van like a lattice-work bulldozer blade, established the height of incoming aircraft. The other was on the roof and rotated at a steady fifteen times a minute, the motors making a low hum. Perched above this roof aerial was a long horizontal bar. When the main aerial picked up an aircraft, the bar transmitted a signal 'interrogating' a device in the plane called a transponder. If the aircraft was friendly then its transponder replied automatically. If there was no reply, or the wrong answer came back, the plane was presumed hostile. The system was called 'identification friend or foe' and it operated instantaneously.

Before the van had been transported here by helicopter, a mound of earth and rock had been raised to make a platform, and it now stood perched on this snowy hump with one set of cables running up to it from a throbbing mobile generator and others for transmitting data snaking down over the edge of the Plata Berget to the control centre on the airfield far below. Nearby was a hut, designed basically for technicians to use. There were two of them here, monitoring the automatic operation of the radar and able to talk direct to the control centre.

When Makarov reached the site in the morning he had been able to lay out a defensive position rapidly. The emplacements which his men had built from blocks of snow resulted from careful prior consideration of how to establish a killing ground into which any attacker coming overland would have to move. A 2-metre-thick wall of packed snow would stop ordinary small arms fire. If he had been permitted to make the emplacements in advance, he could have increased the effectiveness by using frozen snow and water. But that had been forbidden since satellite photography would have revealed it as an obvious military preparation. However, the hut could hold secrets and he had previously laid in a store of ammunition, fuel, food, a powerful 82-mm mortar and several man-portable Strela anti-aircraft missiles. Happily, too, the Spetsnaz were equipped with the new 5.45-mm AK 74 assault rifle, a less clumsy weapon than the old AK 47, which could be slung around the neck for firing on the move and held forty rounds in its plastic magazine.

Better still, now that an attack was imminent, he had the comfort of the fighting patrol in position to cut off the enemy's retreat. So he waited confidently, knowing from the forecast that the cloud would clear during the night.

The hut had a stove and Makarov had organized his twelve men on a roster, two hours on, four hours off, so that there were always eight of them resting. He wanted to keep his tiny force as alert as possible. So far as he was concerned, the two technicians did not count. They were Air Force personnel, seconded from the Air Defence Command, and if the action developed into hand-to-hand fighting, they would merely be in the way. However, General Stolypin had ordained that keeping the radar's surveillance functioning was vital and Makarov had accepted their presence, knowing the decision was a correct one. He was in the hut himself, considering writing a letter to his wife since there was nothing left to prepare, when one of the technicians appeared, letting a blast of freezing air in through the doorway.

'Close that door!' Makarov shouted. 'Were you born in a cow barn?'

'Sorry, Comrade Colonel.' The technician, like most of his breed, had a more casual attitude than disciplined soldiers. 'You're wanted on the blower. The Comrade General's at the Control Centre.'

Makarov put on his fur cap, adjusted the hood of his white camouflage suit to cover it, pulled on his mittens and followed the man out. The snow between the hut and the radar was already pounded down to a path. He hurried across, up the rough steps cut into the mound, and mounted the short ladder to the door.

The interior of the van had the usual custom-built consoles and circular illuminated screens. Although at the moment all the data was passing through to the Control Centre, it could operate on its own. It was warm, sophisticated, pervaded by the gentle hum of its equipment, altogether out of context in this desolate world. God in heaven, Makarov thought, not for the first time, the Air Force have life soft. He sat down heavily on one of the swivel chairs and picked up the telephone.

'Makarov?' Stolypin sounded concerned, which was unusual. 'We have intercepted radio signals from that trawler advising the enemy to cancel the operation. We must assume they will not now attack. What is your fighting patrol reporting?'

'No enemy contact up to 1700 hours, Comrade General. The tracks they were following have been obliterated by fresh snow.'

There was a pause while Stolypin considered this. 'We have more trouble down here. Is it possible that they are heading for Longyearbyen?'

This was the first time Makarov had been asked for advice in weeks. He gave a careful reply. 'Possible, Comrade General, but I think most unlikely.'

'Could you move to intercept them?'

Makarov put down the instrument briefly and went to the small window set into the metal door. 'The visibility remains poor,' he said firmly, 'around fifty metres. I prefer my plan of trapping them between the fighting patrol and here until the weather clears. We could otherwise pass them without knowing.'

'Very well. I will post guards in the valley below the glacier in case they descend.' Stolypin's voice regained its normal curtness. 'Your orders, Comrade Colonel, are to take the enemy alive. You will shoot to kill only if the radar is threatened or in extreme circumstances in self-defence. Is that understood?'

'It is, Comrade General.' Makarov guessed Stolypin had an officer witnessing this conversation and he replied with stiff correctness.

'The continued operation of the radar is paramount and the capture of the enemy of almost equal importance. If necessary you must accept casualties in order to achieve these objectives. You will act in accordance with the prevailing circumstances.'

After Stolypin had rung off, Makarov returned to the hut, furious at the General's final, catch-all, sentence. 'Hell's teeth,' he growled to himself, 'that bastard used always to lecture me that signals reaching Moscow from the West are either too confusing or too politically ambitious to be acted on. He should try responding to his own damn orders.' Ambitious was a mild word for what Stolypin was demanding, and again he had deposited the onus for success or failure on his subordinates, though this time with an added indication that their own lives were of little consequence.

Makarov consoled himself by writing letters both to his wife and their young son, whom it seemed possible he might never see again. 'My dearest Alexei,' he began, 'how are your studies proceeding? Successfully, I hope. Here we are . . .'

Suddenly he tossed the sheet aside. What was this garbage he was writing? The things he wanted to say were simpler – and too dangerous to put on paper. 'Never trust the system. Keep your head. Above all, Alexei my son, try to be a man. Our Russia is ruled by *apparatchiks* like Stolypin, but in the end, in the last counting of heads, you will be respected if you show yourself to have been a man because at heart our people are a great and honest people. Live true to your own spirit and at least you can die content.'

But anyway, how could he say that to a boy just coming up to twelve? He retrieved the sheet, added some perfunctory remarks, signed it and folded it carefully away in an inside pocket. If he was unlucky enough to be killed, someone would find it there.

Soon after six, with the wind whirling flurries of loose snow along the mountainside, Peterson estimated they were at the LUP. He ordered the men to dig in immediately, two to a snow hole, and then eat. Johnson laid out the radio aerial on the snow and prepared to transmit their position. Once they were ready to rest, Peterson intended to get his boots off and do something about his toes.

First, however, he gritted his teeth and padded across on his snowshoes to where Smith was digging. It was essential to confirm the attack plan before they slept, though the cloud fog he needed showed no sign of disappearing. In some ways he would like to have mounted the attack straightaway, but it would be insane. Any enemy would be having an evening meal and be alert, while his own men were not completely bushed, but pretty tired. In seven hours' time the situation would be reversed.

'So that's clear?' he concluded after a few minutes, speaking in a very subdued voice. 'We stay here until 0100, then move across the plateau in two groups.'

'I understand, sir.' If there was residual resentment in Howard Smith's features, it was overlaid by fatigue. Not that Peterson could see much of his face, partly masked as it was by the parka hood and the goggles.

'I want total surprise. No one will open fire until I order, right?'

Smith nodded, but Peterson knew he had to give these men the licence to use their discretion which their skills had earned.

'Unless circumstances make it essential,' he added. 'Like we get bounced.'

'Sure, Colonel,' Smith said wearily. 'The men have all got the message.'

They were interrupted by Johnson, who came padding up softly.

'Sir, I've punched out the signal but I'm getting no reply. There's still a hell of a lot of interference.'

Peterson suppressed a curse. This was not the moment to show his concern. He had to keep morale rising, though he did not relish making a final decision if there was still no confirmation during the night.

'OK, Johnson,' he said. 'Let's just get the snow hole dug, fix some chow and try later.'

While the rest of the team did the same, he helped Johnson dig their hole and, when it was finished, let him boil water for their rations while he also melted snow in his own cook tin. He guessed he was going to need it once he had at long last eased off his ski-march boots.

The moment he had his feet free he knew what was wrong. The wet had got in and later his socks had frozen. He peeled them off and felt his toes. The smaller ones were all right. But the two big toes were a purple red, icy to the touch and totally numb. One way to warm frostbitten feet was to lie with them in your buddy's armpits. But Johnson was busy. A doctor would put the whole foot in tepid water. He waited for the snow to become water in the cook tin and dipped the toes of his right foot in it. At first he felt nothing. Then, as the flesh unfroze, the pain began.

Johnson was worrying unashamedly about the radio. He could operate it through a long lead to the aerial laid on the snow outside and before they ate he sent the message again. This time a light flashed indicating a reply. He had hardly started decoding it with the one-time pad when he looked across at Peterson with unconcealed agitation.

'It's an Op Immediate, sir.'

Next to a 'Flash', an 'Operational Immediate' was the most urgent signal.

Johnson finished the transcription and handed across the pad. It ran: 'Virginia Ridge cancelled. Contact South Wind for exfiltration. Enemy patrols reported your area. Acknowledge.'

'Shit,' Peterson said, the pain in his right foot forgotten.

'Sounds like someone's trying to save our ass,' Johnson commented. South Wind was the codename for the ship whose helicopters would lift them out. 'Isn't that one hell of a damn word-change?'

Peterson did not reply. For how long had Bodo been trying to get through to them? A day, two days? Those brief words spelled extreme danger. Nothing less would have made Admiral King decide to abandon destruction of the radar when they were so close. He squatted beside Johnson and with a stub of pencil printed a reply.

'Your 161835 Op Immediate received. Wind Force Four. Visibility 100 metres.'

'Send that,' he said. 'And get the PRC 77 warmed up. Just as soon as the weather clears enough we'll call in the air.' He didn't want to transmit yet because it would give away his position. Bodo could be relied on to pass the present situation to the ship by secure means.

'And Johnson.'

'Sir?'

'Go tell the others we'll be pulling out and tell Neilson to wake us as soon as the visibility gets above 200 metres.'

Neilson had been allocated the first period of sentry duty. He would alert Burckhardt in the next hole by yanking on a length of line between them, the end of which was attached to Burckhardt's leg. He, in turn, would wake Trevinski by the same primitive, but silent and foolproof method.

Johnson crawled out to convey the orders and Peterson reverted to the problem of his frostbite, thinking sardonically that this was a pretty downbeat end to an operation: waiting with frozen feet to be evacuated before they'd fired a shot. Virginia Ridge was proving no kind of a successor to the Marines' honours, to the tradition of Yorktown and Bellean Wood and Pork Chop Hill. It wasn't going to make even a footnote in the official Defense Department histories. Smith would miss out on his medal. In fact the whole damn thing was the original busted flush. And to cap it all he was going to lose his toenails at the least; the pain as the flesh unfroze was terrible, like being stabbed with needles.

For a while Peterson fought off the agony by rehearsing in his mind how he would call the chopper in, using the strobe light to attract the pilot's attention, finally loosing off a smoke grenade if that failed. The snag with smoke was it drew the whole world's attention to you, though if the wind was only light it would give a measure of concealment too. He wondered what armament the suspected patrols might have. When he had done all he could with speculation, he unpacked a spare pair of socks, checked they were dry, and put one on. Then he repeated the excruciating process with the other foot, all the time thinking through the new tactical problem and wondering when the weather would clear. It must be changing or they would not have achieved the radio contact.

The snow holes were laid out close together, but providing all-round observation. Trevinski, as demolitions expert, had already been teamed up with Smith, and the Dragon missiles were with them. Peterson reckoned that if covering fire was needed for the withdrawal, they could provide it. The Dragon ought to deter any patrol for the necessary few minutes. But the chopper pilots would have to be pretty damned slick.

In the other snow holes the men cooked their meals in pairs on the buddy-buddy system, stacked their equipment handy and then tried to sleep, resentful and mystified at the last-minute word-change.

'It's OK for those pogues in the rear,' Burckhardt complained to Millar, his pride already dented at having had to swop duties with Johnson. 'They can sit back there shooting the breeze and going home to cold beer and hot broads. The only ass-kicking they do is strictly night work. I'd just like to have them out here, sitting in the snow and trying to figure what the goddamn signal really means.'

'I can tell you that for free,' Millar said, his rough-cut accent steadily mocking. 'It means when this cloud lifts we're going to be as safe as a bunch of canaries in a cattery.'

'The Navy advise us to clear the area.' Paulsen turned down the volume of the *Northern Light*'s ship's radio. 'For our own safety.'

'The hell they do,' Clifford commented. 'Which way, huh? The kind of luck we've had can't last.'

It was two hours short of midnight and they had long passed out of the Isfjord's bottleneck entrance, diesels thudding. Weston's instinct, since they were still flying the

Norwegian flag, was to stay within the 12-mile limit. Both he and Clifford had realized that the Russians could be playing cat and mouse with them and it would be far too risky for them to rescue Peterson and the team, even if they were instructed to. On the other hand, it seemed to be inviting trouble to head south for safety, since the Barents Sea was the focal point of the naval confrontation. The only information they had received from the Norwegian Navy, in response to their earlier signal, was that the blockade began at midnight GMT: in three hours' time. If it resulted in the sinking of the Russian merchantman, they would be in deep peril.

'I reckon our best course,' Weston said, deliberating over the charts, 'is to go the opposite of the obvious direction. We ought to head north towards the pack ice.' He tapped his forefinger on the map, where the Svalbard coast was shadowed by a 50-mile-long island, the Karl Prins Forland.

'That's the channel cruise ships take,' Paulsen said. 'Very spectacular scenery.'

'The hell with the scenery –' Clifford began.

'No, no!' Weston interrupted. 'What I want is an inlet among steep mountains.' He indicated a minor fjord, some 20 miles up the channel. 'Can we hide there?'

'In the St Jonsfjord? I think yes,' Paulsen said, then anticipated the next question. 'We can be there in less than three hours.'

'Off we go then.' Weston grinned delightedly. He could try out for real what he had practised several months ago, switching everything off and lying doggo close to cliffs that would disguise the *Northern Light*'s radar echo. 'Won't save us from aerial photography, but maybe they won't look up there. We'll just tuck ourselves away and sit out the next day or two.'

The tugging of the string woke Peterson. He crawled to the snow hole opening. Outside the cloud was clearing fast, shreds of it clinging to the mountaintop like dispersing fog. He roused Johnson, told him to call in the helicopters, and heaved his way out fully, his feet hurting with every step. He had not taken his boots off in case of emergency and he was glad he had not. The agony would have been worse putting them on again. The most important thing now was to confirm their exact location.

His first impression was good. The snow holes would be difficult to spot because they were dug into a drift on more or less flat snow. He looked behind them. The high ridge towered up there to the south. But it seemed farther off than he had expected. He looked west. There was the yawning void of the Longyearbyen valley. He checked with his compass and realized with a sick feeling what was wrong. They were a full 1000 metres farther out on the Plata Berget than he had intended. In the mist they had mistaken a slight rise in the ground for the bottom of the main ridge. He studied his map, then called softly to Johnson, who was now out of the snow hole with the PRC 77 set, its stubby aerial erect.

'Tell South Wind our location is seventy eight fifteen north, fifteen thirty east.'

As Peterson gave the revised position the last tendrils of cloud swept away from the centre of the Plata Berget and revealed the radar. The van appeared to be on a mound. Nearby was a hut, its roof laden with snow. He dropped flat and felt for his binoculars. What were those other shapes? Snow scooters. Soldiers' heads were visible behind snow emplacements. Jesus Christ, the site was defended. They had a mortar too; the black end of its tube was protruding above one of the snow walls. There was a soldier in white, standing up now in the open. Presumably an officer. He was holding a megaphone. Figures rose beside the mortar and bent over the barrel, then were hidden by a puff of smoke. He took a hasty compass bearing, then seized the microphone from Johnson, who was crouching beside him, and transmitted himself.

'This is Six. We have enemy positions azimuth three one

two degrees 1500 metres. Expedite that air, will you? It's going to get hot around here.' They were less than a mile from the Russians and sitting ducks.

Makarov stood unafraid in the centre of his defensive position, bellowing commands. He estimated the range as 1600 metres and deliberately reduced it in his orders because all he wanted to do was keep them pinned down. He shouted to his signaller to contact the fighting patrol. They should be across the glacier by now.

'Comrade Colonel,' the signaller called out when he had finished. 'Their estimated position is 6000 metres from us.'

Even with the weather clear, that would be an hour's ski run. Makarov took the microphone himself, asked for the officer and gave orders to engage the enemy as soon as possible and cut off their retreat. He was working out the most effective way to give fire support to the patrol when he saw two men rise up from the enemy's snow holes and start skiing with furious energy towards a low hump at the side of the plateau. Even without binoculars he could see that one was carrying a missile launcher. He ordered the mortar crew to establish a ranging bracket on the hump. He couldn't pussyfoot any longer. The radar van was not armoured. Any kind of missile could put it out of action. This time the mortar would fire to kill and Stolypin would have to get what propaganda he could from dead bodies.

One of the technicians came running from the radar, stumbling as his boots sank into the snow. 'Two aircraft approaching from the west, Comrade Colonel,' he panted, trying to salute at the same time. 'Range 35 kilometres, speed 170. Control identifies them as hostile. Probably helicopters.'

'Thank you, Comrade.' He dismissed the technician gruffly, despising the man's near panic. His mathematical mind was sorting the speeds and distances almost as rapidly as a computer. All the facts – the timing, the missile, the helicopters, probably gun ships – coalesced into an inescapable conclusion. He was facing a well-coordinated hit-and-run raid of considerable daring. In fourteen minutes or less it could all be over. The fighting patrol was useless. There was no time to call up reinforcements from Longyearbyen. He would have to send some of his own small force on snow scooters to outflank the enemy, while the mortar pounded them and his own Strela missiles dealt with the aircraft. He had barely enough men for all three functions.

Incongruously, at this moment the sun came out, low above the northern horizon behind him, illuminating the enemy more sharply. They would have this midnight sun in their eyes, dazzling them. It was a late bonus, possibly too late.

The first four mortar bombs arched downwards with a whistling scream, and like Peterson the whole team threw themselves down, praying they would not suffer a direct hit. One after the other, at ten-second intervals, the explosions threw up cascades of snow a few yards ahead, momentarily deafening them, while slivers of metal whined past and the snow fell back in a cloud of particles. Then there was silence.

'That was close,' Johnson commented. 'Next time they won't miss.'

'They shouldn't have then,' Peterson snapped. 'Now you get on that set and warn the air we're under fire.' He stood upright, looking for Smith, and waved to him to come across. The captain ran over, sinking through the crust only once, and crouched down.

'Howard, we have to take that mortar out. I don't know what the hell they're playing at with it. Didn't even stop to change the aim.'

'Want to keep our heads down while they get round the flank, I guess.'

'Could be. We still have twelve minutes before the chopper comes. I want you and Trevinski to use the Dragon, right?' He pointed over the plateau to where there was a slight depression running towards the side and a mound of snow. 'That's about the only cover there is within range. I want you over there like you're chasing the devil, while we hold off any flanking move. If you can't get back, I'll have the chopper lift you out from there. You hit that mortar, OK? Not the radar. The operation's cancelled and it stays that way.'

'Sure.' Smith doubled back and almost immediately set off with Trevinski, carrying only the missiles and their sub-machine-guns. The distance was about 600 metres and they skied like racers, leaning forward, arms and legs swinging in the rhythm of a sprint.

Millar watched, lying in the snow, the sniper rifle which was his special weapon in his hand. He reckoned Smith should cover the distance in five or six minutes, saw bombs burst around the hump before they got there, then observed new activity near the radar. Eight Russian soldiers were leaping onto snow scooters. He blinked as they sped off in a flurry of snow. The midnight sun low over the northern horizon was shining straight in his eyes and also reflecting off the snow. He hoisted the sniper rifle into his shoulder and twisted his body round, following the scooters through the telescopic sight. To get around behind the team they would have to pass no farther than 500 yards away. It was the extreme range for accurate shooting and they were travelling fast. He might get a body shot. He would have to gauge the deflection. He shifted farther round, swinging the barrel. The leading scooter turned towards him. He could see the driver's helmeted head. The scooter ran into a shallow gully. When it re-emerged he fired three shots. The driver fell forward, the scooter swerved out of control and the second man jumped clear just before it overturned.

'One down and three to go,' he thought jubilantly. But the other scooters stopped, the men dismounted and ran outwards to form an extended line. Then they began to crawl forward.

Smith and Trevinski knew they would be under fire. The moment they reached the hump they crouched behind it, panting and shaking from their effort. Within seconds another bomb whined down, exploding just the other side, throwing up a curtain of snow. Fragments of metal and splintered rock sang above their heads.

'We have to fix that son of a bitch.' Smith unclipped his skis, positioned himself lying down close to the rock, unfolded the Dragon launcher's bipod, fumbled briefly with the optical sight, then balanced the end of the tube on his shoulder. 'Load!' he yelled. Trevinski thrust in the first round and he squeezed the firing mechanism just as the distant crew in the mortar pit prepared another bomb.

A gout of flame streamed back from the Dragon. For a few seconds Smith could actually see the projectile, a wavering black spot streaking away from him. The snow wall of the mortar pit burst apart as though struck by a hammer, there was a bright flash and he thought he saw a man's arms flung up. More flame blossomed and black smoke plumed upward. A mortar bomb must have gone off too. He heard the two detonations with satisfaction, then a more urgent noise made him shrink back against the rock. The last bomb the Russians had fired was coming down, its shriek growing louder. The white world around him exploded, pain stabbed his lower legs and the launcher was torn from his grasp. He lay, half stunned, until the silence was broken by a moan. He forced himself into a sitting position. Trevinski had collapsed spreadeagled, a huge red stain spreading in the snow. One of his boots was two yards away, his ankle still projecting grotesquely from the black leather. His other leg was obviously shattered, too.

Smith stripped a field dressing off his web gear, where he kept it taped to a strap, and tried to fix it over the stump of the leg. But there was no way it would stop the blood pumping out, so instead he used its fastenings to make a tourniquet below the knee, then ripped open a torn sleeve and jabbed a morphine capsule into Trevinski's arm. The man would probably die of shock, if not of the wounds. He could at least kill the pain.

'You're going to be all right,' he said in Trevinski's ear. 'The chopper'll be here any time. You're going to be right fine.'

Trevinski's head nodded a fraction in acknowledgment and Smith left him to ready another missile round. To hell with not busting that radar, it would be tracking the helicopter anyway, those bastards had it coming. This was the moment he'd planned for and he wasn't letting it slip. His own legs were bleeding profusely. He ignored them and loaded the launcher. It was a slower process single-handed. Then he shouldered it, balanced the bipod on the rock, got the squat shape of the van centred in the sight and fired.

'Get some, sweetheart!' he shouted to no one, as the Dragon jumped and the flame seared back. He felt the heat of it through the parka and concentrated on the guidance.

The impact was magnificent. The van was knocked half off its perch and caught fire, the scanner stopping. A man leaped from the door, at least Smith supposed it was a door. He couldn't see that much detail. In half a minute the radar was blazing. He saw other soldiers running from the hut, one carrying something long. 'Christ, another fucking missile.' He backed off and hastily prepared another round. He had four more left. He'd fix the hut, which must be an ammunition store, and keep a couple in reserve.

From the snow holes, Peterson saw the radar erupt in flame and cursed Smith's impetuosity. Why in hell couldn't he understand that so long as the installation was intact, some Russians had to guard it? Now they were all freed for hot pursuit. The remaining seven from the snow scooters were problem enough, advancing in short runs, then dropping flat again and firing. Neilson, Burckhardt and Millar were fully occupied holding them off. He wanted Mydland available to guide in the choppers. He told Johnson to help ward off the immediate attack and took over the radio himself.

'This is Six. Be informed we have enemy 300 yards east of our position. Repeat east. You want the strobe?'

'That's affirmative,' came the pilot's voice. 'Give us a marker in three minutes. What kind of armament's the enemy got?'

'We knocked out the mortar. They got mainly rifles.'

'We'll be with you.'

Peterson looked up. Low above the mountains lining the fjord were two distinct shapes coming towards him.

'We got the air coming in,' he shouted to the others. 'Three minutes.'

He tried the strobe light. It wasn't working. He'd have to use smoke.

A shot pinged past him and he crouched lower. Goddamn it, this was going to be a close-run thing. And the way the wind was blowing, the smoke wouldn't help either. It could hide them from the choppers, not the enemy.

Action always clarified Makarov's thinking. He stood firm, his white-hooded oversuit blotched with smuts from the fire, calmly assessing the situation. True, he had taken cover when he saw the flash from the missile launcher. Only a fool would have done otherwise. But no sooner had the radar been hit, sparks fizzling from the severed power cables, than he had modified his plan. The mortar lay damaged in its snow pit, one of the crew dead beside it. The next priority was to shoot down the helicopters. Of the three men left, only one was skilled in operating the Strela. He sent him to the hut and the soldier returned seconds before the building itself was hit and began to burn.

'Take up positions,' Makarov ordered calmly. 'I want to get the first helicopter. Fire on my order.'

The soldier went to one of the emplacements and half knelt behind the snow wall, the launcher on his shoulder.

'Keep down!' Makarov roared. He could not bet on the enemy missile firer giving up, even though the helicopters were approaching. So he sent the remaining two Spetsnaz off on skis straight towards the hump. If they could get within rifle range, they could put a stop to that bastard with the missiles. He had observed through binoculars that only one man was operating the weapon. As for himself, he had no option save to wait. He was a poor skier. The most useful function he could perform was to give directions over the radio. The flanking party east of the enemy was closing in very satisfactorily. A surge of hope made him smile. All was not lost. He might yet gain the prizes the Politburo wanted.

Makarov identified the incoming helicopters skirting the contours of the high ridge. A puff of yellow smoke rose near the enemy, becoming a wide stream. He could imagine the belching canister melting the snow. Happily the smoke was blowing back towards the ridge. The helicopters descended perhaps 100 metres apart, the first already tail down as its pilot eased into the hover, preparatory to landing.

'Fire!' he shouted, praying the man's basic aim was good. So long as it was, success would be certain. The missile worked on a heat-seeking principle. Once near the aircraft it automatically sought out the nearest source of heat, namely the engine.

Fifteen seconds later he was rewarded. The helicopter lurched in the air, tipped sideways and crashed. He held the radio microphone to his mouth.

'Assault now. Go in and get them!'

The seven men from the scooters rose and began running awkwardly towards the enemy. The trap was sprung. Makarov watched through his binoculars, hoping for victory. He knew it was his own last chance.

The billowing yellow marker smoke was caught by a shift in the wind and began drifting between Mydland and the helicopters. He had abandoned skis for snowshoes and padded hastily through the acrid fumes to wave the first machine down. To his surprise it was a Norwegian Sea

King, the national roundel clearly painted on its side, the door already open. As the pilot brought it up short into the hover, sending the loose surface snow whirling upwards, there was a massive explosion. He saw the fat body torn open; the machine teetered, shaking more and more violently, then fell out of the air as the rotor blades broke under the strain. For a moment he thought it would land upright, but it tilted and crashed nose first. In the cockpit one pilot raised his hands to protect his face. There was a terrible noise of rending metal, hissing and cracking, and it rolled over sideways.

Mydland tried to run forward, found his snowshoes in the way as he reached the helicopter, pulled them off and climbed up on to the wreck. Inside the cabin the crew chief was groping around, dazed but alive.

'Here!' he shouted, reaching out. The crew chief grasped his hand and he heaved with all his strength as the man tried to struggle out of the doorway, kicking his legs, the dangling intercom lead from his helmet catching on a protrusion. As they scrambled off, there was a further explosion. Heat seared their backs. He pulled the crew chief through the snow, then turned. The wreck was erupting into an inferno of smoke and flame. He searched with his eyes for any sign of the two pilots, the fire scorching his face, forcing him to back away. He prayed that the impact had killed them.

Through the mingling of the black and yellow smoke, he saw the other helicopter approaching. It passed overhead and he hustled the crew chief after it.

Peterson knew what the explosion meant, but kept talking to the American pilot of the other machine, hidden from him by the smoke.

'OK. Come on in. We have missiles that will have fixed that bastard. But there are enemy with rifles pretty close. You have five-cal. machine guns?'

'Sure thing. I'm coming on through.' If the pilot was nervous he wasn't showing it.

'We have four men 50 yards right ahead of you now.'

Peterson was deafened by the chopper's engines as the long, dark green shape loomed above him. He glimpsed the comforting words MARINE CORPS on its fuselage. 'I want you to lift them first.'

The helicopter swivelled in the air, hardly 10 feet up, and he heard the sharp rattle of the machine guns, mounted in a Perspex blister on the body.

'We have your guys visual and also the goons,' came the pilot's voice.

'Great. Lift them and come back for me.'

Out on the snowfield the four men were grouped close together and dug in behind a rough snow bank which they had thrown up with their entrenching tools. Neilson was wounded and Johnson's shooting impaired by his bandaged wrist. Millar glanced back as he heard the whump whump whump of the rotors and saw the American chopper approaching. He wasn't betting any money on getting out of this unhurt himself. Five of the original eight Russian soldiers were still resolutely advancing, crawling through the snow. A head was looking up. He loosed off and it ducked down again. Jesus, that was 100 yards away at most.

'Let's go!' Burckhardt was shouting. Millar looked round again. The helicopter pilot was gesturing at them, thumb up, the order unmistakable. He stumbled across to Neilson, who had been shot in the thigh. 'Hang on to me.' He raised the American in a fireman's lift over his shoulder and retreated, shots whining past him as the Russians rose, but they were impeded by their skis just as he had been. The helicopter was hovering a foot up, its machine gun thudding out bullets, which kicked up spurts of snow around the attackers. Together he and Burckhardt bundled Neilson into the waiting hands of the crew, Johnson tumbled in and Millar felt himself yanked up by the armpits and deposited on the floor.

The helicopter lifted again and clattered sideways. A moment later Peterson appeared beneath, waving at them, his voice all but inaudible.

'We can't leave the gear.'

Millar jumped down and helped throw up skis and packs. A few shots slammed through the fuselage and Johnson cried out in pain. Then they were aboard again and the helicopter lurched away, creating its own snowstorm, the pilot flying behind the smoke of the burning Sea King for cover, coming down again quickly to rescue Mydland and the Norwegian crew chief.

Peterson went forward, climbed up between the pilots, and explained about Smith and Trevinski.

'Sure,' the captain said. 'We'll get them.' He could see that whatever the Russians halfway across were doing, it did not include aiming a missile and the relief was written in his face. As for the ones with rifles, they were no longer a threat. 'Shit, what the hell!' he exclaimed, seeing something else. Distinctly outlined against the snow, away by the radar, a single figure was standing, watching them through binoculars. 'Crazy goon. Who's he think he is, Napoleon?'

Smith was sheltering behind the rocky hump, the Dragon on his shoulder, Trevinski still lying in the bloodstained snow behind him. As the helicopter hovered, Smith fired the sixth and last missile at the advancing Russians, still 600 metres off. But they threw themselves flat and it churned a furrow in the snow before detonating harmlessly.

Peterson leaped down. 'Let's get the hell out!' he shouted angrily, recognizing that Smith's last stand was no longer logical, but the working out of his personal wish fulfilment. He ought to be saving Trevinski, not worrying about Russians who were far out of small-arms range. 'Get going, man!' he yelled. 'Move!'

Reluctantly, Smith wriggled back and helped lift Trevinski into the chopper. They loaded the Dragon and the rucksacks, the crew pulled them back on board, and the CH-46 swayed up and away from the Plata Berget to race close along the high ridge, gaining protection from the mountains on the way to the sea.

Inside the cabin, the crew chief and the gunner fixed a

stretcher for Trevinski. They pillowed his head with a folded combat jacket and Peterson knelt beside him, knowing it would have taken an iron will to survive the shock and the cold this far.

'You'll be fine,' he said, holding the Pole's hand as he was given another shot of morphine and Buckhardt began cutting the blood-soaked and frozen trousers from what remained of his legs. Johnson was occupied with a flesh wound of his own.

Trevinski lay with his eyes open, his black-stubbled skin seeming grey. He managed a few words, almost too faint to catch over the reverberation of the engines.

'Are they OK?'

'Sure, everyone's fine. We made it!'

'Yeah, but . . .' The effort was costing him a lot and suddenly Peterson understood. It was only in movies that dying men ask if their side had won. What Trevinski wanted to know was more basic. When his legs got shattered, what had happened further up?

'You betcha,' he said, reassuring him with all the emphasis he could muster. 'They're just fine. No problem. Only reason you can't feel too much is that the medic gave you another shot. You'll be all right, friend, and the broads just better watch out when they see you coming.'

Trevinski's mouth twitched in what might have become a grin and he closed his eyes. It was only when they had touched down on the ship that Peterson discovered he had died.

The sailors who welcomed them, crowding the deck, were both jubilant and curious. Peterson felt he might have dropped in from another world. He understood their amazement better when he caught sight of himself in a mirror in the captain's cabin. His eyes were dark-circled, his skin pallid and his face haggard. His overwhites were filthy and stained with blood, though what they could not see was the worst, his toes. He told the captain he would need medical attention.

'We'll have all of you checked over. Is there anything else you want?'

'Could I call London, England?'

'I think we could fix that.' The ship's communications connected with Washington in seconds. A call could be patched through easily. 'You do know it's long after midnight there?'

'I suppose it is.' Peterson laughed. 'I don't think she'll mind. At least I can tell her it's all over. Not that she knows what it is, of course.'

The captain glanced at him sharply. Was it possible the Delta Force had been operating in such isolation that they didn't know about the blockade?

'Well,' he said cautiously, 'your piece of the action may be over. Ours could be about to start.'

The clock on his cabin wall showed 2351 GMT.

Nancy Peterson had slept better and woken on Friday morning feeling more in control of herself than she had for several days. She had breakfasted late, then gone down to the hall of the apartment block to see if there was any mail. She thought her sister might have written and expected nothing from Tom. She knew he would always pick up a phone rather than put pen to paper.

Each apartment had its own numbered mailbox and the hall porter had reached into hers as she approached the desk. He was an absurdity, the hall porter, with a florid face and stiffly waxed moustachios, wearing a canary yellow uniform adorned with much incongruous gold braid. He looked as if he had stepped out of a period play.

'Here you are, madam,' he had announced in his slightly high-pitched voice. 'And the very best of the morning to you, madam.' He had held out a single airmail envelope and stood unashamedly watching her.

The moment she had seen the Norwegian stamp and postmark she felt a surge of excitement. It was from Tom.

He really must be changing his ways and he must be all right if he was in Oslo. She had torn it open and the few words had shrieked at her that everything was not all right at all. Everything was all wrong.

'Be back in a few days. Take care. I love you. Tom.'

All her recently found confidence ebbed away.

'Can I be of assistance, madam?' The hall porter had leaned across the polished counter of his private enclave, curiosity rampant in every syllable of his words, his red-veined eyes intent on her face. He lived for the petty dramas of the apartment block, encouraging confidences from the lady tenants. 'I trust you have not received bad news, madam?'

'No, no. Nothing's the matter,' she had stuttered, thrusting the letter into her handbag, and had stumbled out onto the street, not even thinking where she was going, knowing that she didn't want to talk to anyone, least of all him. She felt lonelier than she could ever have imagined.

She walked down the street and bought a *Herald Tribune* from a corner newsvendor. Its headlines offered only gloom. 'Intransigent Soviet Mood. Blockade deadline midnight.'

To make things worse, it was a beautiful summer day, the leaves on the plane trees fresh and green and dappled with sunlight. On an impulse she decided to walk to Hyde Park, anything was better than waiting at home.

The twenty-minute walk had calmed her down a little and once in the park she had sat on a wooden bench and began to read the newspaper seriously. A short while later a middle-aged Englishwoman had joined her on the same bench and spoke to her after she put the paper down.

'Tell me, what do Americans feel about this?' She had sounded educated and sympathetic.

'Oh!' The question had caught Nancy unawares. How did this stranger know she was an American anyway? Then she realized how completely her clothes gave her away, the crisp shirtwaist dress must be enough by itself. What did

Americans think? she had asked herself. Editorial comments reprinted in the *Herald Tribune* were full of righteous indignation both against abuse of Norwegian sovereignty and strategic threats to the United States. But the farmers in Kansas, or at home in Minnesota, did they care? To her parents and their neighbours, world events mattered only if they affected the price of gas for their tractors. Would even sophisticated New Yorkers back the risks of military action? In honesty she could not believe they would. Yet if you elected leaders, you had to trust their judgment. That was what she had decided yesterday.

'Ours is a big country. Most Americans wouldn't want to fight over anything if they could avoid it,' she had explained. 'But I guess there are times when we just have to.'

'I think so too,' the woman had said. 'My son's in the Army. He was in the Falklands.'

'Those islands must have seemed pretty far away.' Nancy remembered Britain's undeclared war with Argentina in 1982.

'Eight thousand miles. But the inhabitants were British. This Norwegian place isn't terribly close either, is it? Though wars do start over the most extraordinary things. I'm only praying my son won't be sent there.' As if reading Nancy's mind she had added, 'In the end one's just glad if they come back alive.'

The Englishwoman's common-sense, if slightly fatalistic, attitude had cheered Nancy considerably. At last she had met someone in the same predicament, or almost the same. They had gone for coffee together and exchanged addresses before they parted, promising to call each other soon.

For the rest of the day she killed time as best she could, going to a movie in the afternoon, always horribly aware that at midnight GMT, which was 1 o'clock in the morning by British Summer Time, war would either be about to start or have been averted.

The late evening TV news had merely confirmed, with much military and naval detail, how intense the Great

Power confrontation had become. The Russian merchant ship carrying the radar masts was already close to the southern coast of Svalbard.

Nancy had gone to bed and just managed to doze off, or so it seemed, when the phone rang. She fumbled for the bedside light and snatched the instrument from its cradle.

'May I speak with Mrs Nancy Peterson, please?' The voice was American and sounded very distant.

'Nancy Peterson speaking.' She began to tremble.

'Hold the line, please. I have an official call for you.' There was an echo of the words, then silence.

Oh God, she thought, waiting in agony, it's to say he's dead.

His heaviness of spirit mirrored in the ponderous gait of his skiing, Makarov trudged across the killing ground and rallied his remaining men. Behind him the radar and the hut smouldered, sending a funereal wrack of smoke into the crystal arctic air. The transformation since yesterday was extraordinary. The sun had risen farther into the sky, although it was still not 1 a.m., bathing the ranges of mountains in picture-postcard glory. There were snow-laden peaks as far as he could see. He was on top of a white world and he cursed it, feeling a sudden hatred for the extreme of climate, for the sleepless summer and the forced hibernation of winter dark. If ever there was a place the capitalists deserved to keep, by God this was it. Svalbard was worse than Siberia.

When Makarov reached the enemy's laying-up point he was out of breath and his feet ached. Aided by three unwounded soldiers, he began a painstaking search for evidence. Around the snow holes they picked up a pair of Norwegian-made *langrenn* skis, a glove and empty cartridge cases of various calibres. But he knew none of these was sufficient. What Stolypin wanted was an American in American uniform, with identity tags and letters from home. He had wanted one alive. The latest order said one dead would be adequate. So he laboured across to the helicopter they had shot down, hoping against hope.

Though the aluminium skin had burned to a silvery ash, the ribs of the machine were only partially collapsed, like one of those whale skeletons by the shore. The long rotor blades, bent nearly double by the impact, would establish to the experts that he had correctly identified this as a Sea King. The question was: whose? Several NATO navies had them. He found part of the tail intact, with numerals still decipherable on it, and knew with something like despair in his heart that this Sea King had been Norwegian. It was not what the General required. The Norwegian government could invent a hundred reasons for the helicopter's legitimate presence over Svalbard. So he spared his men the unpleasant task of digging for the charred corpses of the crew. Those could wait. They would keep well up here, damn it.

After he had given a brief report over the radio, Makarov stood stamping his feet for warmth, until the helicopter came from Longyearbyen. Unless the Politburo discerned some propaganda value in making accusations without evidence, and he was unable to conceive how they could, this whole episode would cease to have happened. The radar would have burned due to an electrical fault, the dead have been killed in accidents. The people of Russia would never know the firefight on the Plata Berget had taken place. But within the Party and in the KGB Secretariat it would be known, and he did not think it a recipe for promotion. He was glad that he had warned Anya in advance. At least she might understand if his next job was not even in the Army, and they lost their housing privileges. No one was unemployed in the Soviet Union, but he had gloomy visions of what the bureaucracy might devise to occupy him.

The Norwegian captain of the destroyer had heard about

the lost Sea King and anger stiffened his resolve. He had been shadowing the Russian merchantman since yesterday afternoon, waiting both for the midnight deadline and for her to enter the 12-mile limit around Svalbard. There was so much disputed water in the Barents Sea that his Prime Minister had settled for an interpretation of territorial rights which no nation could challenge. At 3 a.m. on Saturday the radar showed the ship as approaching the Sorrkapp, the southernmost tip of Svalbard. The captain ordered full speed ahead and the destroyer surged into the heavy swell, the sea breaking over her bows and tons of foaming water hammering her decks.

At ten cables' distance the captain radioed on the international frequency and warned the merchantman that she was infringing Norwegian limits. When an equivocal answer came back, he sent his crew to action stations, deliberately allowing the order to be broadcast. Was the Russian slowing? He watched her long hull, noting the extensive aerials above her bridge, unable to conceal his tenseness from the officer of the watch. Behind them, dispersed in a thousand square miles of the Arctic Ocean, stood considerable reserves of naval power. But backing the merchantman was the whole Soviet Northern Fleet and its air forces. If he opened fire, his own ship was certain to be the next casualty. Yet if the Norwegian Navy was not prepared to fight for Norwegian sovereignty, there was little point in its existence. He took the microphone again and warned the Russian in both Norwegian and English that if she did not stop he would open fire. Then he gave range and bearing to his guns.

There was no reply. The captain was about to give a final warning, when the merchantman began to turn to starboard, heeling over slightly. The destroyer followed her round, keeping the guns trained on her all the time, the barrels rising and dipping to compensate for the swell, until she had swung through 180 degrees and was heading south. When she eventually left the territorial limit the captain took a strong pull of acquavit from a silver flask his wife had given him and then passed it to the officer of the watch.

'Skol,' he said. 'You should drink to the luckiest day of your life.'

'Would she have had a political officer on board?' the young lieutenant asked. 'Directing the captain, I mean?'

'No question about it.' The captain screwed back the flask's polished cap. 'You saw those aerials? They were probably getting orders from Moscow.'

He watched the merchantman through his binoculars, gradually going hull down towards the horizon, and wondered how far the incident really had been a test of nerve. It seemed more probable that some private international bargain had been struck beforehand. And yet . . . how could one tell with the Russians? They had been deliberately provoking Norway's resolve for years before the Svalbard crisis, sending submarines into the mainland fjords, defying regulations. The only thing he knew for certain was that they respected strength and that the crisis might now be over.

After he had showered and changed into fresh fatigues, Peterson sat down to write his official report. He had his feet in a bowl of cold water and they hurt like hell. The doctor reckoned he would lose his toenails. But whatever the pain, he wanted to get things on paper right away, before time blurred the details. One thing was certain. He was not recommending Howard Smith for any decoration, nor anyone else. All they had carried out was a withdrawal from a place they had never officially been in, though as he described the action in formal phrases he knew instinctively that the Pentagon would end up giving someone the big medal. If word of Operation Virginia Ridge ever leaked out to the American public, and he did not see how it could be kept quiet, what with the Freedom of Information Act, the episode would need a hero to outweigh the casualties.

When he thought about it, Smith was bound to be the

man. Held off the enemy single-handed and all the rest of the crap. The truth was unlikely to be revealed. The last three days, which had seemed like three weeks, were already sliding into history and once they were history, Jesus, they were going to read like something quite different. In fact, recalling the captain's puzzlement, it occurred to him that maybe he didn't know so much about it all himself. Maybe he never would.

He was halfway through when the phone buzzed.

'Sorry about the delay, sir,' came the nasal voice from the ship's communications centre. 'We were kind of busy with priority traffic. I have Mrs Nancy Peterson on the line now, sir.'

'Is that you, sweetheart?' He paused, knowing this call was going halfway round the world and back.

'Oh, Tom. I can't believe it!' She was unable to hold in her emotion, didn't want to. 'Oh, Tom, is that really you? I thought they were going to say you were dead.'

'Me?' He made the idea sound as crazy as he could. 'Whatever made you think that? I'll be back with you in a couple of days, sooner maybe.'

'Where are you, darling?'

'On a Navy ship.'

'Tom, I was so worried. Is it all over?'

'Over? Sure it's over.' He allowed himself a white lie. 'The Russian boat just turned around a few minutes ago and headed back home. Like we're going to. Hey, Nancy, let me tell you something.'

'Yes.' She hung on his words, breathless and apprehensive, fearful of whatever duty he was being called for next.

'I love you, sweetheart.'

When she put the phone down in the cramped London apartment, Nancy had ceased to care whether he left the Marine Corps or not. Later, perhaps, she would return to that subject. Right now, it was enough that he was safe. What had that English lady in the park said? 'In the end one's just glad if they come back alive.'

10

Frederik Folvik inspected the control tower with careful concern. Longyearbyen airfield had been the last place the Russians relinquished in their progressive withdrawal. It had taken them a month, less two days, to remove whatever they had installed in the so-called Aeroflot hangar on the other side of their runway. The building was merely a shell now, though the new Aeroflot manager, Simonov, made a lot of fuss about its security. Aeroflot was back in its original offices, in the building next to the tower. Folvik shook his head in private wonderment at the Russians' cool resumption of past arrangements. He knew that the handover ceremonies of the day before yesterday had depended on the Soviet Foreign Minister and the United States Secretary of State initialling the heads of agreement for a new Arctic Treaty. More immediately he wished he had been here himself to check the air-traffic control equipment.

The Negretti and Zambra aneroid barometer seemed unharmed. He tapped the glass, then polished the chromium with a corner of his handkerchief. The long array of radios had already been checked by technicians. They would be getting an approach radar soon. That was good. In fact, many things were good. He had been through a training course and received promotion, in anticipation of the radar installation. His gaze strayed up the Plata Berget. All the snow had melted. Stark though the mountain looked, he knew the warmth of July would have brought out tiny flowers in the valleys, probably on the summit too. He wondered if the new radar would be put there. He must go up himself one day. People said that there was still a burned-out hut, though the Air Force had kept silent about other wreckage they had removed. The Plata Berget seemed likely to keep its secrets.

Folvik continued his measured appraisal of his little domain, to which he had never expected to return, until his eye was caught by something definitely out of place.

Among the aviation circulars on the noticeboard was a newspaper cutting. He put on his glasses to read it and blinked with amazement. Someone had pinned up a newspaper photograph of Colonel Makarov. He was in an immaculate uniform, a row of medals slanting along the lapels of his jacket. Facing him was an officer with considerably more insignia and braid who was pinning a further decoration on Makarov's chest. The Colonel's expression was staunchly proud. Underneath was a lengthy caption in Russian.

Well, Folvik thought, I've something better for the notice board than that, anyway. He opened his briefcase and replaced the cutting with a proper photograph, signed along the bottom. 'FF. With love from us both.' It was Annie's wedding picture. She looked lovely. He was glad for Paul Mydland and also sorry. Svalbard wouldn't be the same without Annie, though she had been due to leave soon in any case.

Later it occurred to him not to throw away the Russian cutting. He was curious about Makarov. So he clumped down the stairs and sought out Simonov for a translation.

The Aeroflot manager and his staff were circumspect and polite, reflecting the official restoration of friendly relations between the two countries. Simonov read the caption earnestly and explained.

'This is the ceremony honouring the Comrade Colonel after his promotion to Major General. The medal is for outstanding service to the people. He will now go as Chief of Staff to Tanna-Tuva.'

'To where?'

'Not a place you would know, my friend.' Simonov smiled with the self-satisfaction of a man who understood the subtle gradations of promotion in his own country. 'Tanna-Tuva is a military garrison on the Chinese border. Remote, but of great strategic importance.'

'Must be cold there,' Folvik observed.

'Very. But generals are privileged. They do not feel the cold.' He smiled again. 'None the less, if I were him, I would prefer to be nearer Moscow. If there is a war with China, that is where it will start.'

Folvik thanked him, mystified at what paradox had earned Makarov promotion to such a hellish-sounding place. Then he returned to the tower. There was an SAS flight due in half an hour. When the pilot called, he was about to answer 'Runway is free for landing' as he always had, when he remembered. He had been promoted himself. He had more authority.

'You are clear to land,' he said firmly. A new era in his life had begun. This evening he would eat the new season's cloudberries his mother had sent. With cream.